SpringerBriefs in Cell Biology

More information about this series at http://www.springer.com/series/10708

Pedro F. Oliveira · Marco G. Alves

Sertoli Cell Metabolism and Spermatogenesis

 Springer

Pedro F. Oliveira
CICS—UBI—Health Sciences Research
  Centre
University of Beira Interior
Covilhã
Portugal

Marco G. Alves
CICS—UBI—Health Sciences Research
  Centre
University of Beira Interior
Covilhã
Portugal

SpringerBriefs in Cell Biology
ISBN 978-3-319-19790-6   ISBN 978-3-319-19791-3   (eBook)
DOI 10.1007/978-3-319-19791-3

Library of Congress Control Number: 2015950895

Springer Cham Heidelberg New York Dordrecht London

Printed on acid-free paper

Springer International Publishing AG Switzerland is part of Springer Science+Business Media
(www.springer.com)

Acknowledgments

The authors wish to acknowledge Ana D. Martins, Raquel L. Bernardino, Tânia R. Dias, Tito T. Jesus and Luís Rato for their contribution to the final version of this book.

Contents

Chapter 1
Introductory Remarks

In the 19th century Enrico Sertoli described for the first time the existence of large and irregularly shaped testicular somatic cells that later would be known as Sertoli cells. It was subsequently reported that adjacent Sertoli cells form intercellular connections that enable the establishment of the blood-testis barrier, a key structure for the formation of viable spermatozoa in a process named spermatogenesis. Thus, the proper maturation, proliferation and function of Sertoli cells are closely associated with the male reproductive potential. These cells, known as "nurse cells", are essential for the normal development of germ cells by offering not only physical support, but also for creating an immune-privileged environment and nutritional support for the developing germ cells. The presence of Sertoli cells promotes the establishment of the appropriate microenvironment that allows the occurrence of a normal spermatogenesis.

Spermatogenesis maintenance in vivo is highly dependent on the metabolic cooperation established between Sertoli cells and developing germ cells. Sertoli cells metabolism primarily depends upon extracellular glucose and presents some particular features, as the majority of this substrate is converted to lactate, instead of being oxidized in the Krebs cycle. Moreover, developing germ cells, particularly pachytene spermatocytes and round spermatids, are unable to metabolize glucose and use the lactate produced by Sertoli cells as an energy substrate. Recent studies reported that Sertoli cells are capable to adapt their glucose transport machinery and metabolism to ensure the appropriate production of lactate for the occurrence of spermatogenesis, even in conditions of glucose limitation. In addition, it has been suggested that the production of lactate can also be derived from aminoacids or the metabolism of glycogen, although these processes are not entirely understood.

© The Author(s) 2015   1
P.F. Oliveira and M.G. Alves, *Sertoli Cell Metabolism and Spermatogenesis*,
SpringerBriefs in Cell Biology, DOI 10.1007/978-3-319-19791-3_1

For many years the metabolic cooperation established between testicular cells was disregarded, but recent advances have highlighted the relevance of these processes for male fertility. It has been proposed that the reproductive outcomes of several pathological conditions known to impair the male reproductive function, such as Diabetes Mellitus, may be due to changes in the metabolic cooperation between Sertoli cells and the developing germ cells. Thus, the understanding of the functioning and regulation of these metabolic processes is a crucial step to identify key mechanisms associated with Sertoli cell (dys)function, and to enlighten their influence over spermatogenesis and male (in)fertility.

Chapter 2
The Sertoli Cell at a Glance

The testicles are compartmentalized organs whose functional units, the seminiferous tubules, are immersed in a web of loose connective tissue and interstitial cells (or Leydig cells). In the seminiferous tubules, the epithelium is compartmentalized due to the presence of unique somatic cells, named Sertoli cells (Fig. 2.1). These cells function as the structural elements of the seminiferous epithelium, physically supporting the ongoing of spermatogenesis and regulating the flow of nutrients, growth factors and other substances to male germ cells. In mammals, at the time of puberty, Sertoli cells suffer a profound alteration on their morphology and function, becoming morphologically and biochemically distinct from the undifferentiated cells.

2.1 Sertoli Cell Structure and Morphology

When fully differentiated, the Sertoli cell is a columnar shaped epithelial cell of large dimensions, which extends from the base of the seminiferous epithelium, the basement membrane, to the lumen of the tubules (Fig. 2.1) (Brooks 2007; Foley 2001; Rodriguez-Sosa and Dobrinski 2009). The basal portion of the Sertoli cells adhere to the basement membrane (or basal lamina), a fibrous structure composed of various extracellular matrix proteins (such as laminin, collagen and heparan sulfate) that maintains the structural integrity of the seminiferous tubules (Hadley et al. 1985).

© The Author(s) 2015
P.F. Oliveira and M.G. Alves, *Sertoli Cell Metabolism and Spermatogenesis*,
SpringerBriefs in Cell Biology, DOI 10.1007/978-3-319-19791-3_2

In mammals, the size of the Sertoli cell shows a high variation, with the individual volume of the cell ranging from approximately 2000–7000 μm^3 (Russell and Peterson 1984; Russell et al. 1990). The volume density of these cells (expressed as a percentage of the entire seminiferous tubular epithelium) also varies within the several mammalian species, from approximately 15 % in mice to 40 % in humans (Russell et al. 1990).

Electron micrographs showed that the Sertoli cell possesses an expanded and clear cytoplasm with an irregular shape that varies from species to species (Russell and Brinster 1996; Russell et al. 1996). The majority of the Sertoli cell surface is in close association with the surface of germ cells (Fig. 2.1), illustrating the extent to which the Sertoli cell expands its cytoplasm to directly interact with the developing germ cells. In the cytoplasm, these cells exhibit several organelles, in particular great quantities of mitochondria, which indicate a high metabolic activity (Russell 1993a).

In most species, an irregular shaped nucleus can be found in the basal portion of the cytoplasm of the Sertoli cell. This nucleus is of large dimensions (up to 850 μm^3), with an irregular shape that depends on the stage of the seminiferous

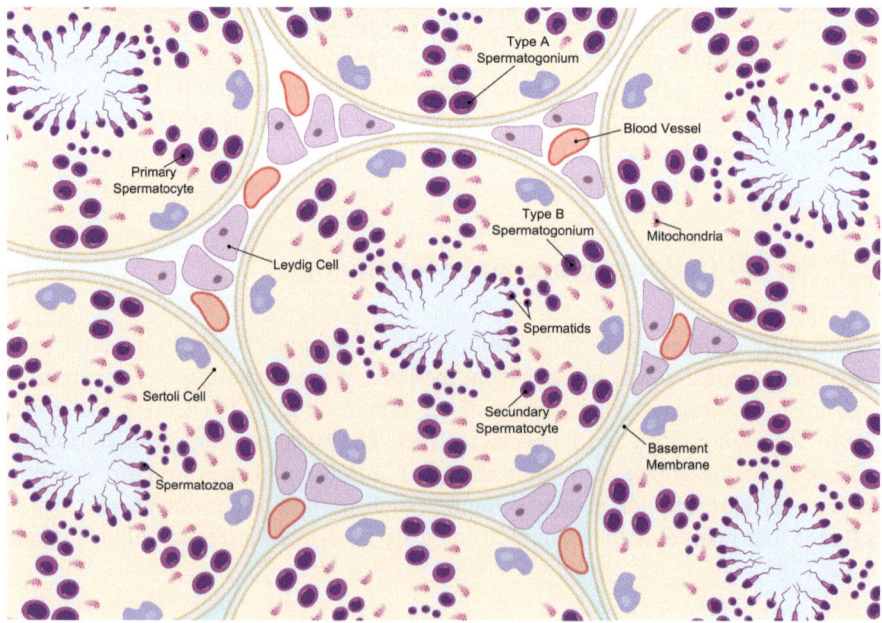

Fig. 2.1 Schematic illustration of the seminiferous tubules. The seminiferous epithelium is formed by Sertoli cells and germ cells at different stages of development. Sertoli cell extends their cytoplasm from the basement membrane to the lumen of the seminiferous tubule. Leydig cells and blood vessels can be found in the interstitium, which is filled by the interstitial fluid. The majority of the surface of the Sertoli cell is in close association to several germ cells. In spermatogenesis, diploid spermatogonial stem cells origin male haploid germ cells from, through cellular division and differentiation

tubule cycle (Russell et al. 1990; Ye et al. 1993) and on the age of development of the individual (Heyn et al. 2001; Krzanowska and Bilinska 2000). The nuclear envelope is invested with deep invaginations (or indentations) that are associated with an accumulation of vimentin intermediate filaments. Moreover, multiple nuclear pores can be found in the nuclear envelope, whose density depends also on the stage of the spermatogenic cycle (Cavicchia et al. 1998). Another characteristic of the Sertoli cell nucleus is the large dimension nucleolus with a three-partite structure that can easily be recognized in the nucleoplasm, usually with two chromocenters associated at diametrically opposed sides in which the centromeric regions of the chromosomes are clustered (Guttenbach et al. 1996; Kushida et al. 1993).

An extensive network of continuous tubular structures with a few ribosomes attached to its surface, known as endoplasmic reticulum (ER), can also be found in the cytoplasm of the Sertoli cell. In the basal portion of these cells, the ER is associated with tight junctions (Brokelmann 1963; Dym and Fawcett 1970; Flickinger and Fawcett 1967) and ectoplasmic specializations (Russell 1977a; Russell and Clermont 1976), being part of the Sertoli—germ cell junctional complexes. While the rough ER is sparse and can be found in the basal region of the cell, the smooth ER is abundant in adult Sertoli cells and can be found near mitochondria, indicating that these cells are intensely involved in the metabolism of lipids or steroids (Russell 1993a). Mitochondria are quite numerous in Sertoli cells and dispersed among the other organelles. Their shape is very variable, with spherical or elongated mitochondria being more abundant, but they can also be cup- or donut-shaped (Bizarro et al. 2003). Tubular cristae are predominant, but the foliate forms are also present in mitochondria of Sertoli cells (Bizarro et al. 2003). These cells exhibit a small Golgi apparatus, being that this single network of stacked saccules is normally dispersed in the supranuclear region of the cell (Chen and Yates 1975; Hess and França 2005; Rambourg et al. 1974, 1979). Lysosomes and multivesicular bodies, resulting from the phagocytotic activity of Sertoli cells, are situated throughout the cytoplasm of the cell (Fig. 2.2), predominantly around the residual bodies and ectoplasmic specializations, with their abundance depending on the stage of spermatogenesis (Hess and França 2005).

The ectoplasmic specializations are cell junctional complexes that are established between two Sertoli cells (Russell 1993b) and between Sertoli cells and germ cells (Vogl et al. 2000), relying on cytoskeletal elements for structure and function, mainly actin filaments, but also microtubules and vimentin (Vogl et al. 1993). The adjacent Sertoli cells are also linked by tight junctions, creating a tight barrier known as the blood-testis barrier or Sertoli cell barrier. This junctional structure has a porosity of approximately 1000 daltons, ensuring that nothing with a higher molecular weight passes to the lumen of the seminiferous tubule (Dirami et al. 1991; Fawcett et al. 1973; Foley 2001). Besides being associated with elements of the cytoskeleton and with the smooth endoplasmic reticulum (McGinley et al. 1979; Russell and Clermont 1976; Russell 1977a, b, 1979a, 1993a), the ectoplasmic specializations and tight junctions have been linked with several other Sertoli cell proteins, including occludin, espin, sertolin and gelsolin (Li et al. 2014;

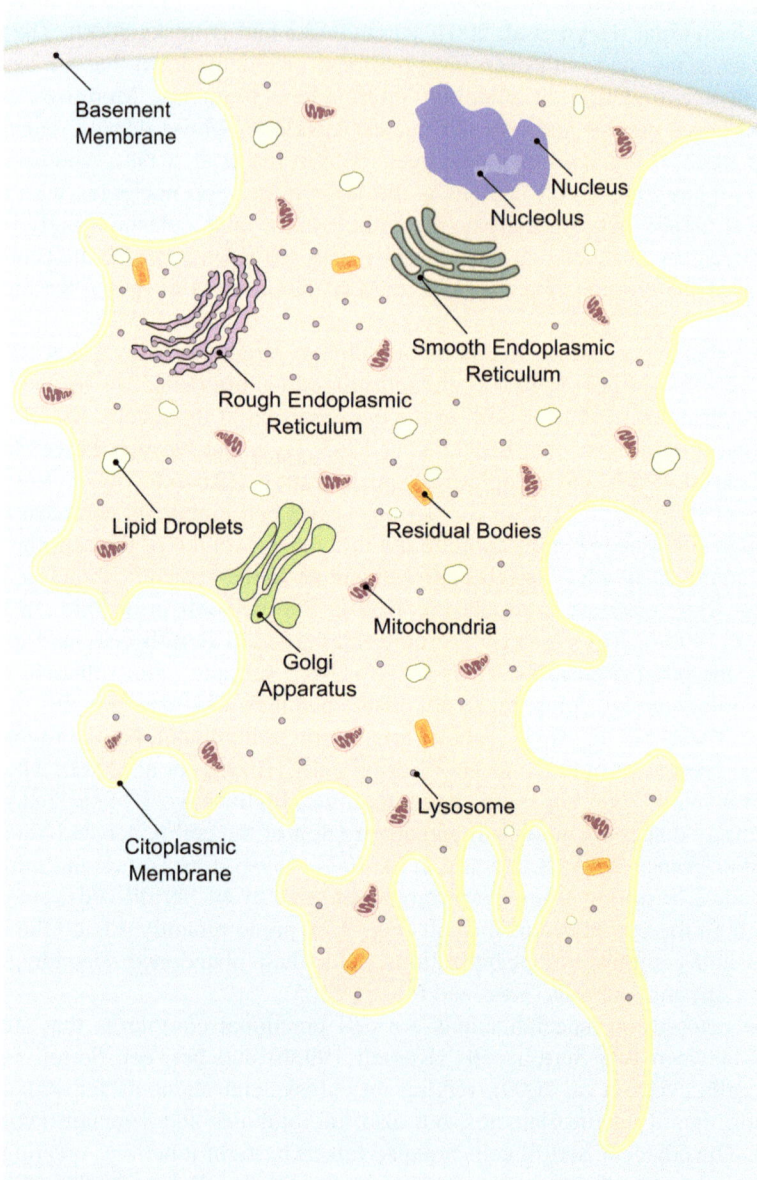

Fig. 2.2 Diagram of a Sertoli cell. The Sertoli cell is a columnar shaped epithelial cell of large dimensions, where the basal portion of the cell adheres to the basement membrane. The mitochondria are abundant and disperse on the cytoplasm. The nucleus envelope presents deep invaginations and a large dimension nucleolus with a three-partite structure. The rough endoplasmic reticulum is sparse and can be found in the basal region, while the smooth endoplasmic reticulum can be found near mitochondria. The Golgi apparatus is normally dispersed in the supranuclear region of the cell. Lipid droplets and lysosomes are disperse throughout the cytoplasm of the cell. The lysosomes can be found around the residual bodies

Cheng and Mruk 2002). However, the organization of these junctional complexes is variable from species to species (Moroi et al. 1998; Parreira et al. 2002).

Several other components have been observed in the cytoplasm of Sertoli cell, among which lipid droplets and glycogen particles (Guo et al. 2003; Sertoli 1865; Slaughter and Means 1983), typically in the basal compartment, with their quantities varying between species and according to the spermatogenesis stage (Russell 1993a; Erkan and Sousa 2002; Fouquet 1968). The presence of lipid droplets has been associated with the ability of Sertoli cells to recycle lipids from the breakdown of germ cell degeneration and from residual body phagocytosis, although this hypothesis is not entirely consensual. Glycogen particles and glycogen metabolism-associated enzymes have also been described in Sertoli cells (Slaughter and Means 1983). The amount of glycogen present in these cells is stage and species dependent and an increase of glycogen storage has been reported in several pathological conditions (e.g. Sertoli cell only syndrome and Sertoli cell tumors) (Anniballo et al. 2000; Henley et al. 2002).

2.2 Sertoli Cell Physiology

The Sertoli cell is the key somatic element responsible for the regulation of spermatogenesis and for the establishment of the different rates of spermatozoa production (Orth et al. 1988; Walker and Cheng 2005). These cells, commonly known as "nurse cells", have many important roles not only in the development of the testicles and its functions, but also in the expression of the male phenotype (Mruk and Cheng 2004; Sharpe et al. 2003). The central functions of the Sertoli cells are: (1) formation of the blood-testis barrier; (2) providing structural and nutritional support to the developing germ cells; (3) phagocytosis of residual bodies and degenerating germ cells; (4) production and release of regulatory factors; (5) establishment of a localized immune-privileged environment (Dym and Raj 1977; Feig et al. 1980; Alexander 1977; Johnson et al. 2008; Jutte et al. 1982; Setchell 1980; Silber 1978).

Among the most important functions of these cells is the creation of the special environment necessary for the development of germ cells and the regulation of the endocrine environment of the seminiferous tubules (Setchell 1980). The formation of the blood-testis barrier is responsible for excluding many of the substances that are present in the interstitial fluids from the tubular lumen (Setchell 1969). The blood-testis barrier tight junctions successfully avoid the movement of large molecules from the interstitial space to the lumen of the seminiferous tubules (Dym and Fawcett 1970; Aoki and Fawcett 1975).

Sertoli cells execute many of their functions by extending the cytoplasm in thin arm-like processes (in two dimensions) and cylindrical or sheet-like processes (in three dimensions), enclosing the multiple developing germ cells. Junctional interactions are established between Sertoli and developing germ cells, particularly ectoplasmic specializations (Russell 1977a; Russell and Clermont 1976), to help connecting these cells. Desmosome-gap junctions are also present at the structural interface between Sertoli and germ cells, to help bind these cells together, being particularly abundant in spermatocytes (Russell et al. 1983).

The seminiferous tubule luminal fluid serves not only as a mean of transport for sperm, as, in addition, it allows the establishment of a favorable microenvironment for the occurrence of spermatogenesis. This fluid begins to be produced by Sertoli cells when they reach their mature state (Sharpe et al. 2003). Its composition is very stable due to the presence of the blood-testis barrier and is very different from the circulating plasma and testicular interstitial fluid. This fluid has been described as rich in Na^+ and Cl^-, with a K^+ concentration twofold higher than that of the other extracellular fluids (Clulow and Jones 2004). Another important feature of this luminal fluid is the control of its pH. This parameter is maintained by the action of intracellular buffers and also by the balance between the production and elimination of protons (Roos and Boron 1981). In fact, the Sertoli cell expresses several types of membrane ion transporters directly involved in the movement of acidic and basic particles through the membrane (Bernardino et al. 2013a). While the role of these membrane transporters is largely unknown, its operation and regulation is essential to determine the osmolarity and pH of the seminiferous tubule luminal fluid (Oliveira et al. 2009a, b). Various ion channels have also been identified in Sertoli cells (Rato et al. 2010). These channels serve as pores that allow the passive diffusion of ions through the membrane of these cells. Ion channels are typically selective for a particular ion species or a limited group of ion transport direction and depends on the electrochemical gradient that is established. These channels are regulated by multiple factors, such as Ca^{2+} concentration or pH (Hubner and Jentsch 2002) and have several functions, as for instance the transduction of electrical and chemical signals, the transepithelial transport, the regulation of cell volume and cytoplasmic or vesicular ion concentration. Furthermore, Sertoli cells must provide the nutritional supply during the development of germ cells (Griswold and McLean 2006). If in the early stages of development the germ cells use glucose as energy source, obtained from the systemic circulation (Riera et al. 2001; Brauchi et al. 2005), in later stages of development, germ cells loose the capacity to metabolize glucose (Boussouar and Benahmed 2004). So, spermatids and spermatocytes are dependent on lactate supplies provided by Sertoli cells (Bajpai et al. 1998b; Jutte et al. 1982; Nakamura et al. 1984a).

Paradoxically, despite being known as sustentacular cells, Sertoli cells are able to induce apoptosis of developing germ cells in an essential regulatory process, taking into account that one mature Sertoli cell can only support a limited number of developing germ cells. Thus, when the cell number exceeds this threshold, it is necessary to eliminate them to allow the spermatogenic process to adequately proceed (Xiong et al. 2009). In addition, the phagocytic activity of the germinal cells components is another important function of Sertoli cells (Griswold et al. 1988). The selective phagocytosis of germ cell residual bodies is a receptor-mediated activity that occurs in Sertoli cells (Morales et al. 1985, 1986). It results from the recognition of specific markers on plasmatic membrane of those residual bodies (Clermont et al. 1987), which have specific antigenic determinants distinct from those present in the membranes of non-degenerating germ cells (Bellve and Moss 1983; Millette and Bellve 1980).

The secretions of the Sertoli cell, particularly growth factors and hormones, are also critical to the control of spermatogenesis and the reproductive function of males (Skinner 2005), since they strictly regulate the maturity of the male gonads and the safeguard of spermatogenesis (Levine et al. 2000; Clermont and Perey 1957). Several proteins and factors are secreted by these cells, such as the androgen binding protein (ABP) (Fritz et al. 1976b), transferrin (Skinner and Griswold 1980), glycoproteins (O'Brien et al. 1993), sulpho-proteins (Elkington and Fritz 1980) and myoinositol (Robinson and Fritz 1979). The Sertoli cell is also responsible for the secretion of several cytokines and other specific products for the development of germ cells (such as the c-Kit ligand, activin and inhibin) (Syed et al. 1988) that influence both germ cell development (Pollanen et al. 1989) and Sertoli cell function (Nehar et al. 1998; Huleihel and Lunenfeld 2002) and whose role will be discussed in the subsequent chapters.

2.3   Hormonal Control of Sertoli Cell Function

Hormones are key regulatory factors of the functioning of the Sertoli cell (Fig. 2.3). Among those, follicle-stimulating hormone (FSH), sex steroid hormones, thyroid hormones (TH) and insulin deserve a special emphasis. FSH is secreted by the pituitary in response to the secretion of gonadotropin releasing hormone (GnRH), which acts through G-protein coupled receptors, that, in the testicles, are exclusively located in the Sertoli cell (McLachlan et al. 2002). FSH has a central role on the male reproductive potential. Individuals with no functional receptors for FSH have smaller testicles. In fact, it has been reported that knockout mice for the receptor of this hormone have a very small number of Sertoli cells (Dierich et al. 1998), apart from presenting an altered spermatogenesis. Moreover they are diagnosed as azoospermic and/or teratozoospermic, although frequently classified as fertile (Tapanainen et al. 1997). This gonadotropin controls the proliferation of Sertoli cells during the perinatal period and the pubertal phase, determining the spermatogenic ability of male adults. The mechanism of action of FSH involves the cyclic AMP signaling pathway, through activation of G-proteins. FSH also increases the levels of phosphorylated protein kinase B (PKB-P) via a phosphatidylinositol 3-kinase (PI3K)-dependent mechanism (Meroni et al. 2002). The PI3K is an enzyme involved in the regulation of various biological processes, including mitogenesis, oxidative stress and glucose metabolism. FSH also stimulates the production and secretion of the hormone inhibin B by Sertoli cells. In fact, plasma levels of inhibin B are clinically used to evaluate the function of Sertoli cells during childhood. On the other hand, during adulthood, inhibin B levels are dependent on the presence of germ cells and therefore are reported to reflect the functional state of spermatogenesis (de Kretser et al. 2004). Of note, FSH also stimulates the production of other factors and substances, such as transferrin, lactate and androgen receptors.

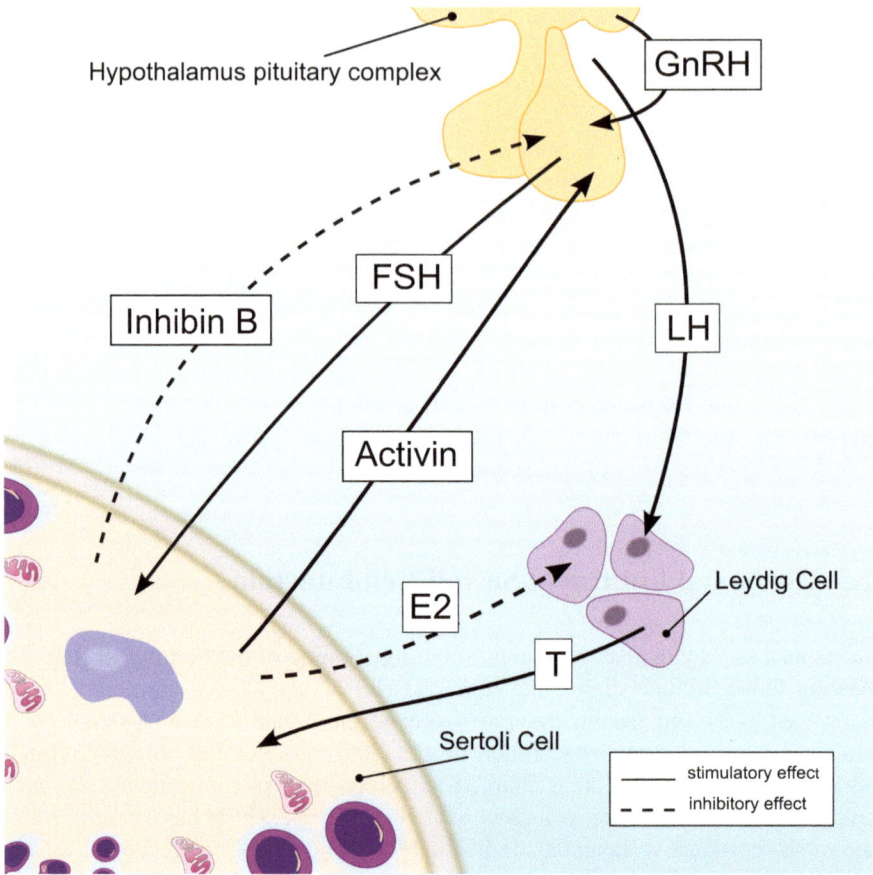

Fig. 2.3 Hormonal regulation of hypothalamus pituitary gonad axis. The gonadotropin releasing hormone (*GnRH*) is synthesized by hypothalamus, which will stimulate the pituitary to produce the luteinizing hormone (*LH*) and the follicle-stimulating hormone (*FSH*). LH binds membrane receptors on Leydig cells and stimulates the testosterone (*T*) production. FSH binds membrane receptors on Sertoli cells, stimulating the production/secretion of 17β-estradiol (*E2*), activin and inhibin B. E2 acts on Leydig cells inhibiting the production of testosterone. Activin and inhibin B produce a positive/negative feedback on pituitary, respectively. Abbreviations: *GnRH* gonadotropin releasing hormone; *LH* luteinizing hormone; *FSH* follicle-stimulating hormone; *T* testosterone; *E2* 17β-estradiol

The secretion of FSH is modulated by prolactin levels. While lower prolactin levels stimulate Sertoli cell growth, lactate production and synthesis of some proteins, elevated prolactin inhibits FSH secretion, which may compromise the male reproductive function (Scarabelli et al. 2003). However, the disruption of the secretion of this prolactin or its receptor in mice does not affect fertility, which illustrates that although this hormone serves as a controlling mechanism of Sertoli cell function, it is not vital for the reproductive function in males.

Spermatogenesis is dependent on the presence of a suitable intratesticular level of sex steroid hormones (androgens and estrogens). The Sertoli cells express androgen receptors, whereas germ cells lack these receptors (Lyon et al. 1975). Androgens are known as the male sex hormones and include various steroids, such as testosterone, androstenediol and 5-α-dihydrotestosterone (DHT). They have a central role on the masculinization of the reproductive tract and genitalia during the sexual differentiation process (Hughes 2001; Sultan et al. 2001) and are essential for initiation and maintenance of spermatogenesis (Roberts and Zirkin 1991). In fact, germ cell development is arrested at the spermatocytes stage (Chang et al. 2004) or early spermatid (De Gendt et al. 2004; Holdcraft and Braun 2004) when androgen receptors are ablated in Sertoli cells. The androgen receptor activity in Sertoli cells is regulated by testosterone (and its derivate DHT) (Lindzey et al. 1994), which is produced by Leydig cells in response to luteinizing hormone (LH). In fact, the physiology of the Sertoli cell is modulated by testosterone metabolites such as DHT, which has a biological activity two to three times higher than testosterone (Robaire and Viger 1995). Both androgens (testosterone and DHT) bind to and activate the same receptors, although DHT presents a much higher affinity (Deslypere et al. 1992). Androgen receptors are essential for normal spermatogenesis and when they are silenced on the Sertoli cell, the development of germ cells halts in the spermatocyte or primary spermatid stages (De Gendt et al. 2004; Holdcraft and Braun 2004), illustrating that the Sertoli cell is the preferred site of action for androgens in the control of spermatogenesis. Furthermore, it has been reported that increased expression of androgen receptor is directly related to the maturation of Sertoli cells (Buzek and Sanborn 1988). Indeed, androgens are necessary for the promotion of the integrity of the blood-testis barrier and assembly of junctional complexes (Meng et al. 2005; Wang et al. 2006).

The role of estrogens in male reproductive function has been under discussion in recent years. It is now accepted that these sex hormones also play an important role in the development and maintenance of the male reproductive function and, consequently, in male fertility (Carreau et al. 2008; Nilsson et al. 2001; O'Donnell et al. 2001). Estrogen biosynthesis is catalyzed by aromatase (O'Donnell et al. 2001). In the testicles, aromatase is expressed in immature Sertoli cells (Abney 1999; Fritz et al. 1976a), the major source of estrogens in prepubertal individuals, while in adults, the Leydig cells are the main responsible for the synthesis of estrogens (Carreau et al. 2009). A high concentration of estrogens has been reported in the testicular interstitial fluid (Rato et al. 2013) and in the *rete testis* (Free and Jaffe 1979). Moreover, estrogen concentration in the rat epididymis is approximately 25 times higher than in plasma (Kumari et al. 1980). These data suggest that estrogens play an active role in the control of spermatogenesis and male reproductive function. In fact, estrogens can modulate apoptotic signaling pathways in rat Sertoli cells (Simões et al. 2013), prevent the development of Leydig cells and inhibit the production of testosterone, stimulating spermatogenesis by decreasing apoptosis in post-meiotic spermatogenic cells and altering the proliferation and differentiation of spermatogonial cells (O'Donnell et al. 2001). Although the role of these hormones in the physiology of Sertoli cell is a controversial topic, their action is mediated through the interaction with their specific receptors. All the subtypes of the membrane-bound (GPR30) and of the nuclear estrogen receptors (ER), ERα and ERβ, were identified in Sertoli cells (Oliveira et al. 2014a; Pelletier and El-Alfy 2000; Saunders et al. 2001, 2002; Taylor and Al-Azzawi 2000). Furthermore, Sertoli cell ER-knockout null mutations cause profound alterations in spermatogenesis and infertility (Chung et al. 1998).

The presence of insulin receptors has also been described in Sertoli cells (Oonk and Grootegoed 1987). Recently, it has been reported that insulin-deprived Sertoli cells present altered expression of metabolism-associated genes involved in the production and export of lactate, as well as altered secretion of lactate and consumption of glucose (Oliveira et al. 2012). The TH have been associated with an inhibition of the Sertoli cell cycle (Walker 2003). Neonatal administration of triiodothyronine (T3) to rats suppresses the proliferation of Sertoli cells (van Haaster et al. 1993). Of note, men with hyperthyroidism are frequently diagnosed with oligospermia and present a considerable lower sperm counts (Clyde et al. 1976). In addition, individuals with thyroid deregulation usually present symptoms of erectile dysfunction and decreased libido or impotence (Wagner et al. 2008). Hypothyroidism has been associated with a prolongation of the growth phase of the Sertoli cells, translating into a delay of the differentiation of these cells. This leads to an increase in the number of Sertoli cells and promotes the enlargement of the testicles and sperm production (van Haaster et al. 1992). Sertoli cells also possess receptors for the growth hormone (Gomez et al. 1998) and a deficiency on this hormone is associated with a smaller size of the testicles, illustrating a reduction in the number of Sertoli cells, although fertility is not compromised (Bartlett et al. 1990).

2.4 Pathophysiology of Sertoli Cells

The success of spermatogenesis is on the basis of male fertility. Taking into account that Sertoli cells support, protect and nourish the developing germ cells, the proper functioning of these somatic cells is essential for male reproductive function. Any failure in the process of maturation of the Sertoli cell means that these cells may no longer be able to support the development of germ cells. This is a dynamic process and the absence of germ cells, such as in cases of exposure to radiation, may lead to severe alterations in the function of the Sertoli cell, prompting the cell into a stage functionally identical to the immature state. Several pathologies are associated with de-differentiation of Sertoli cells. For instance, in seminiferous tubule constituted only by Sertoli cells, these cells exhibit one or more markers of an immature state (such as the anti-mullerian hormone or cytokeratin-18) and may even show a total absence of androgen receptors, which are known to be absent in immature Sertoli cells (Bar-Shira Maymon et al. 2000). Furthermore, Sertoli cells with immature phenotypes have been consistently associated with testicular germ cell cancer incidence, which has more than doubled in the last decades. Cryptorchidism, a condition in which the testicles did not descend properly from the abdominal cavity into the scrotum, is one of the most important risk factor for developing testicular cancer and also a factor directly associated with low production of sperm. Testicles of individuals with cryptorchidism contain seminiferous tubules only with Sertoli cells and these cells have a number of characteristics associated with the immature state. In fact, cryptorchidism has been suggested to be more a consequence of an impaired differentiation of these cells than a cause for this phenomenon (Skakkebaek et al. 2001). In the last decade, a new concept associated with the male reproductive health as arisen: the testicular dysgenesis syndrome. This syndrome comprises several pathological manifestations as hypospadias (congenital malformation of the urethra), reduced sperm counts, testicular cancer, among others. Although these disorders may manifest at different stages of sexual development, they are thought to have a common origin in fetal life as the result (at least in part) of the abnormal functioning of Sertoli cells (Skakkebaek et al. 2001).

The Sertoli cell is also susceptible to the action of many toxic substances, pesticides and heavy metals. Many of these substances affect the cell cytoskeleton or induce chromatin condensation and vacuolization of the cytoplasm. The effect of endocrine disruptors and other environmental pollutants on the physiology and function of the Sertoli cell has received special attention from several researchers. Studies have consistently demonstrated that the administration of the synthetic estrogen diethylstilbestrol to rats reduces the number of Sertoli cells, disrupting the formation and maturation of the blood-testis barrier (Toyama et al. 2001). Recently, it has also been reported that a concentration of 2,4-dichlorophenoxyacetic acid in the range of values that can be found in the urine of men who directly work with this pesticide induces significant changes in the metabolism of Sertoli cells, with putative effects on spermatogenesis (Alves et al. 2013b).

As previously referred, one of the major functions of the Sertoli cell is to ensure the transport and metabolism of glucose to produce the metabolic precursors necessary for the developing germ cells (Robinson and Fritz 1981). Deregulation of the metabolic behavior of these somatic cells can compromise the energy supply to germ cells. Several metabolic diseases, which are currently major threats to public health (where we can highlight Diabetes Mellitus), have their genesis in insulin resistance and/or absence of insulin, as well as an inability of cells to respond efficiently to the stimulation by this hormone. In these circumstances, the metabolism of Sertoli cell undergoes important changes (Alves et al. 2012; Oliveira et al. 2012), causing impairment in the development of germ cells and, thus, in male fertility.

In fact, both the malfunction of Sertoli cells, as disturbances in their differentiation and/or maturation have been identified as crucial factors in the origin of the reproductive dysfunctions associated with various pathological conditions. It is essential to understand the physiology, structure and function of the Sertoli cell, as well as the mechanisms involved in its development and maturation, to better comprehend how all these processes can be influenced by our lifestyle and environment. It is likely that a more detailed knowledge of the functioning and maturation of the Sertoli cell will point toward specific molecular targets to counteract the deleterious effects of pathologies that affect the reproductive health of males.

Chapter 3
Spermatogenesis

Male fertility is highly dependent on the success of spermatogenesis, the multi-step process of male germ cell expansion and development that results in a continuous daily production of millions of spermatozoa (Sharpe 1994). An adult male mouse or rat produces about 10 to 30 million spermatozoa per day, while a fertile man produces more than 40 million spermatozoa per day, beginning at puberty and spanning his entire reproductive life (Cheng and Mruk 2013).

Spermatogenesis is a process controlled by a network of endocrine and other regulatory factors (Rato et al. 2012b; Verhoeven et al. 2010), in which immature germ cells undergo mitotic division, meiosis and differentiation to give rise to the millions of mature sperm cells. This process occurs within the seminiferous tubules, the functional unities of the testicles, by a close association between germ cells and Sertoli cells (O'Donnell et al. 2001; Shubhada et al. 1993; Walker and Cheng 2005). Sertoli cells are the key cellular element supporting spermatogenesis, contributing to the formation of the spermatogonial stem cells niche required to the continuous maintenance of the spermatogenic event, and for the nurturing of developing germ cells (Saunders 2002).

The creation of the blood-testis barrier, by the establishment of specialized adhesion junctions between adjacent Sertoli cells near the basement membrane of the seminiferous tubules, divides the seminiferous tubule into basal and adluminal compartments. During the initial stage of spermatogenesis, developing germ cells must pass through the blood-testis barrier, moving from the basal into the adluminal compartment. Once in the adluminal compartment, the germ cells continue to develop into spermatozoa in a defined and immunoprivileged microenvironment (Smith and Walker 2014).

© The Author(s) 2015
P.F. Oliveira and M.G. Alves, *Sertoli Cell Metabolism and Spermatogenesis*,
SpringerBriefs in Cell Biology, DOI 10.1007/978-3-319-19791-3_3

The process of spermatogenesis is one of delicate complexity, requiring 6–9 weeks for completion, depending on the species. It involves a synchronized series of mitotic and meiotic divisions, elaborated differentiation steps, and constantly changing intercellular interactions. All of these processes are monitored by a remarkable interplay of endocrine, paracrine and autocrine factors (Clermont 1972; Russell et al. 1993). Spermatogenesis involves a stringent balance between self-renewal and differentiation of spermatogonial stem cells to ensure an endless production of mature spermatozoa. During this intricate process, spermatogonial stem cells not only replicate themselves for maintaining the pool of stem cells, but also progressively differentiate into various types of male germ cells, including spermatocytes, and spermatids (Fig. 3.1). At the beginning of spermatogenesis, diploid spermatogonia proliferate producing three populations of cells with markedly different purposes: (1) one subpopulation of spermatogonia that are presumably identical to their progenitors and keeps its function as stem cells; (2) a majority of spermatogonia that enter a differentiating pathway and become spermatozoa; (3) and a great number of spermatogonia that undergo apoptosis. Every step of the spermatogenesis is precisely regulated by the microenvironment or the niche of the testicles. The niche is mainly established by the Sertoli cells and is composed by growth factors, cytokines, and metabolic substrates, among others (Dym 1994; Hofmann 2008).

3.1 Phases of Spermatogenesis

Spermatogenesis is a multi-step process involving three major phases (mitosis, meiosis and spermiogenesis), as well as other cellular events such as cell migration, differentiation and apoptosis. These events are highly regulated and the understanding of the molecular mechanisms that regulate the three phases of spermatogenesis has been a major focus of study for decades (Smith and Walker 2014).

3.1.1 Mitosis

In the mitotic (or proliferative) phase, spermatogonia undergo either self-renewal or differentiation, with both processes involving successive divisions. The spermatogonial proliferation phase occurs in the basal compartment of the seminiferous epithelium. The division of spermatogonial stem cells near the basement membrane of the seminiferous tubule initiates the spermatogenic process.

The number of spermatogonial types that have been identified varies amongst the different mammalian species. In the human male very little information is known about human spermatogonial renewal mechanisms. Still, type A pale, A dark and B spermatogonial forms can be distinguished in the human testicles (Fig. 3.2 Panel I) (Dym 1994). The identification of the exact human

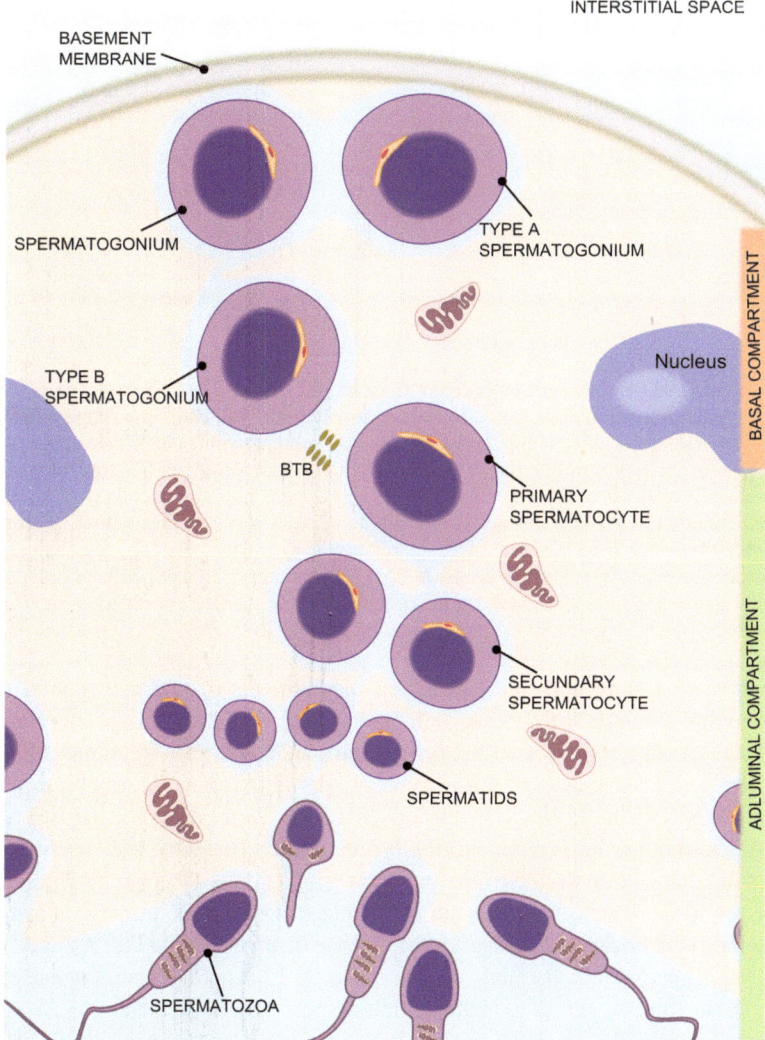

Fig. 3.1 Schematic illustration of the seminiferous epithelium and Sertoli cell. The blood-testis barrier (*BTB*) physically divides the seminiferous epithelium into basal and adluminal compartment. The seminiferous epithelium is composed of Sertoli cells and developing germ cells at different stages. The germ cells form intimate associations with Sertoli cells and a single Sertoli cell is in contact with multiple germ cells. The supporting Sertoli cells adhere to the basement membrane where spermatogonia are also adherent. Spermatogonia type A divide and develop into spermatogonia type B, which enter meiotic prophase and differentiate into primary spermatocytes that undergo meiosis I and form the haploid secondary spermatocytes. Meiosis II yields four equalized spermatids that migrate toward the lumen where fully formed mature spermatozoa are finally released in lumen of seminiferous tubule. Abbreviations: *BTB* blood-testis barrier

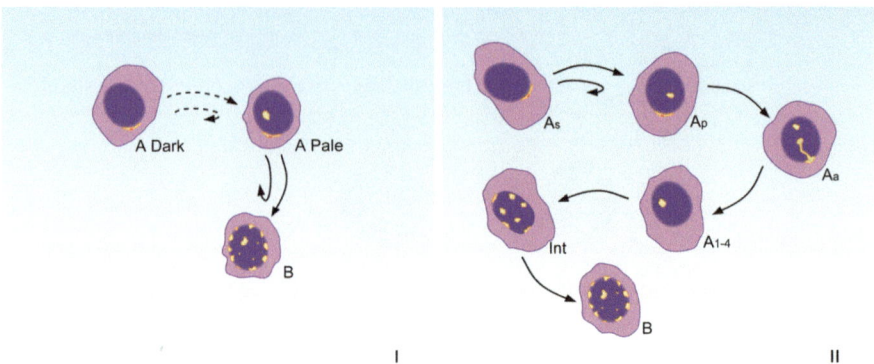

Fig. 3.2 Spermatogonial stem cells renewal in human and rodent testicles. Panel I—Human spermatogonial renewal mechanisms. Human spermatogonial stem cell is a subpopulation of the type A dark and/or A pale spermatogonia. The A pale spermatogonia can divide and either origin new A pale or the more differentiated type B spermatogonia (B). Panel II—Rodent spermatogonial renewal mechanisms. Spermatogonial stem cell divides into single cells (A_s) or into paired spermatogonia (A_p). The A_p spermatogonia divide successively forming type A spermatogonia (A_a), which is subdivided into type A_{1-4} spermatogonia, intermediate (*Int*) spermatogonia, and type *B* spermatogonia (B). Abbreviations: *B* Type B spermatogonia; A_s Spermatogonial stem cell; A_p Paired spermatogonia; A_a Type A spermatogonia; *Int* Intermediate spermatogonia

spermatogonial stem cell and the actual mechanisms of spermatogonial renewal still remains a matter of debate. It is speculated that the human spermatogonial stem cell is a subpopulation of the type A dark and/or A pale spermatogonia.

In the rodent testicles, multiple type A spermatogonia, intermediate and type B spermatogonia have been identified (Fig. 3.2 Panel II). The undifferentiated type A spermatogonia is conventionally considered the only true spermatogonial stem cell, that either divides into two new single cells (A_s) or into paired spermatogonia (A_p). These A_p suffer an incomplete cytokinesis, staying connected by an intercellular bridge (Oakberg 1956; Roosen-Runge 1952). The A_p spermatogonia divide successively forming chains of up to 32 aligned type A spermatogonia (A_a). These three types of undifferentiated spermatogonial cells (A_s, A_p and A_a) are morphologically identical, being indistinguishable by light microscopy on a histologically stained cross-section of testicle. Recent studies suggested that spermatogonia A_p and A_a still possess stem cell characteristics. The spermatogonia A_a further undergo a differentiation step with associated morphological changes and transform into A_1 spermatogonia. Following the formation of A_1 spermatogonia, these cells suffer a series of five mitotic divisions, giving origin successively to A_2, A_3 A_4, intermediate and B spermatogonia. Spermatogonia A_1 to B are considered differentiating spermatogonia and each step of differentiation is associated with a mitotic division, with the exception of the transition from A_a to A_1 spermatogonia (Yoshida et al. 2007). The type B spermatogonia in turn differentiate into prelep- totene spermatocytes, that undergo a final replication of nuclear DNA, and then transit across the blood-testis barrier to enter the apical compartment for cell-cycle progression, until they transform into primary spermatocytes and initiate meiosis (Cheng and Mruk 2013).

3.1.2 Meiosis

The second phase of spermatogenesis, commonly known as meiosis, occurs while the spermatocytes remain intercalated between cytoplasmic processes of adjacent Sertoli cells on the adluminal side of the seminiferous epithelium (Bellve et al. 1977). Germ cells that enter meiosis are called spermatocytes. The early primary spermatocytes are transported through the blood-testis barrier into the adluminal compartment, leaving space in the basal compartment for the next generation of spermatogonia to enter mitosis (Grootegoed et al. 2000). The first meiotic division is subdivided into four stages: leptotene, zygotene, pachytene and diplotene. During the leptotene stage, chromosomes begin to condense, and immediately after, in the zygotene stage pairs of homologous chromosomes are formed. Following, during the pachytene stage, synapses are completed and then substituted by crossing-over and homologous recombination. Finally, in the diplotene stage, chromosomes are unsynapsed and the cell divides (Fig. 3.3) (Kotaja 2013).

Meiosis is characterized by the separation of chromosomes that occurs during the metaphase, anaphase and telophase of the first meiotic division, after which distinct germ cells, named secondary spermatocytes, are originated (Schulz and Miura 2002). These cells quickly enter a second meiotic division of short duration, without a DNA synthesis phase. It is during this second meiotic division that each secondary spermatocyte produces two spermatids genetically unique, with a haploid number of single chromosomes (Kotaja 2013; Schulz and Miura 2002).

3.1.3 Spermiogenesis

Spermiogenesis or haploid differentiation is the final stage of spermatogenesis, which includes a series of dramatic differentiation events that eventually culminate with the development of a highly differentiated spermatozoon (Fig. 3.3) (Russell et al. 1993). A hallmark event of spermiogenesis is the loss of 80–90 % of the cellular and nuclear volume of spermatids (Sprando and Russell 1988). This final phase of spermatogenesis, consists in a complex morphological transformation of the haploid germ cell that leads to the release of late spermatids into the lumen of the seminiferous tubule (Bellve et al. 1977). Spermatids undergo morphological changes such as the establishment of the flagellum, the formation of the acrosome and the elongation of the nucleus (Russell and Peterson 1984). A great portion of the cytoplasm is also eliminated and the chromatin is gradually condensed, with the sequential replacement of the somatic and testicular histones. The histones are replaced by transition proteins, several highly basic proteins and small

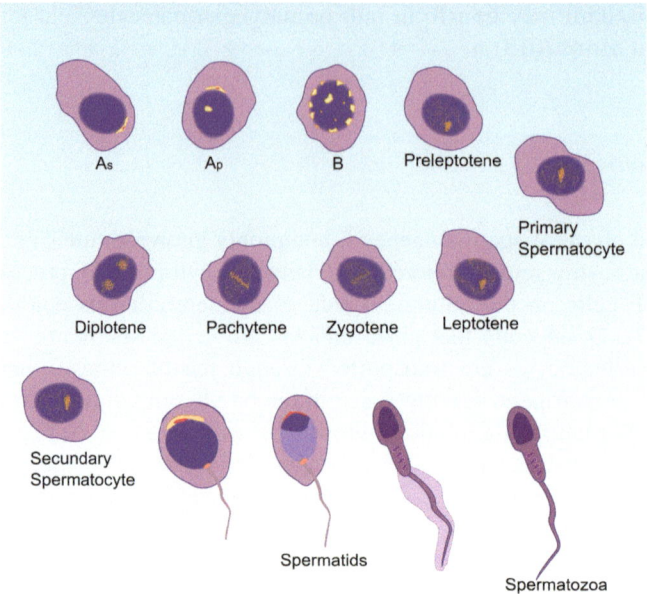

Fig. 3.3 Stages of spermatogenesis. In the mitotic phase, spermatogonia undergo either self-renewal or differentiation, with both processes involving successive divisions (A_s and A_p). In turn, type B spermatogonia (B) differentiate into preleptotene spermatocytes until they transform into primary spermatocytes and initiate meiosis. The first meiotic division is subdivided into four stages: leptotene, zygotene, pachytene and diplotene, after which secondary spermatocytes are originated. These cells enter a second meiotic division and produce two spermatids genetically unique. Spermatids undergo morphological changes and highly differentiated spermatozoa are produced. Abbreviations: A_s Spermatogonial stem cell; A_p Paired spermatogonia; B Type B spermatogonia

lysine-arginine-rich proteins that help in the transformation of the nucleosomal chromatin into smooth condensed chromatin fibers (Hecht 1998; Zini and Agarwal 2011). Towards the end of spermiogenesis, protamines (the key sperm nuclear proteins) replace the transition proteins, while in the nucleus the acrosome (a vesicle containing a group of proteolytic enzymes) develops. Numerous alterations also occur in the cytoplasm of the spermatid leading to the formation of the axoneme and the tail. As a consequence, several testicle specific proteins are synthesized during spermiogenesis (Hecht 1998).

3.1.4 Spermatogenesis and Cycle of the Seminiferous Epithelium

The duration and number of stages in each cycle of spermatogenesis is distinct in the different mammalian species. Germ cells undergo changes through 19, 16 and 6 steps in rats, mice and humans, respectively, and eventually transform into spermatozoa. This fact allowed the division of the seminiferous epithelium into discrete stages (I–XIV, I–XII or I–VI in the rat, mouse or human, respectively), comprised of unique cellular associations between developing germ cells (in particular spermatids) and Sertoli cells. These cellular associations can be observed in cross-sections of seminiferous tubules (Cheng and Mruk 2013; Kotaja 2013; Mruk and Cheng 2004).

Mouse spermatogenesis is divided in 16 developmental steps, with the first 8 steps comprising the round spermatid phase and the last 8 steps the spermatid elongation. Step 1–8 spermatids are present at the stages I–VIII of the seminiferous epithelial cycle, and the initial elongating spermatids, step 9 spermatids, are visible at stage IX. During stages VII–VIII, adhesive connections between Sertoli cells and spermatids are remodeled as the spermatids start to elongate. At this stage of development, desmosome-based connections between Sertoli cells and the less mature germ cells are displaced by the stronger ectoplasmic specialization adhesion complex, which is specific for elongated spermatids. Subsequent steps of elongation are established during stages X–XVI, and the mature step 16 spermatids are finally released into the lumen during spermiation at stage VIII (Kopera et al. 2010; Kotaja 2013). Shortly before spermiation, the Sertoli cells play an active role in reducing the cytoplasmic volume of the elongated spermatids, and the adhesive contacts between the spermatids and Sertoli cells are destroyed (Grootegoed et al. 2000). Spermatozoa are then released into the tubule lumen, so that it continues maturation through the *rete testis*, efferent ducts and epididymis, before acquiring full fertilizing capacity in the female reproductive tract (Cheng and Mruk 2013; Mruk and Cheng 2004).

3.2 Regulation of Spermatogenesis

Spermatogenesis is a strictly regulated process that requires a singular environment created by Sertoli cells. Though this control can arise due to the action of multiple combined factors, it occurs at two major levels: (1) hormonal or endocrine and (2) paracrine or autocrine. A summary of the actions of the hormonal and other key regulatory factors is depicted in the present chapter, while the role of key extrinsic factors produced and delivered by the somatic Sertoli cells on spermatogenesis is discussed in the subsequent chapter.

3.2.1 Hormonal Regulation of Spermatogenesis

The hypothalamic-pituitary gonadal axis is the master hormonal control system that regulates spermatogenesis and all the male reproductive tract (Fig. 2.3) (Schulz and Miura 2002). Within this axis, neurons of the hypothalamus produce

GnRH that enters the hypothalamic-pituitary portal system and binds to receptors on the plasma membranes of pituitary cells. Pulsatile GnRH signals stimulate gonadotroph cells in the anterior pituitary to secrete FSH and LH, that then act on the testicles to regulate the spermatogenic potential (Walker and Cheng 2005). An essential function of LH is to stimulate the production of testosterone by Leydig cells, via activation of the LH receptor (LH-R). Testosterone production influences the development of peritubular and Sertoli cells, and consequently of germ cells. Leydig cells are found amongst the seminiferous tubules, in the interstitial spaces of the testicular tissue, from where testosterone diffuses into the seminiferous tubules, thereby mediating the biological activity of LH. As within the seminiferous tubules, only the Sertoli cell possesses receptors for testosterone and FSH. Thus, these somatic cells are the ultimate targets of the hormonal signals that regulate spermatogenesis. Yet, testosterone also acts through peritubular myoid cells to maintain spermatogenesis, although this effect is believed to be mediated via the Sertoli cells (Welsh et al. 2009).

The second gonadotropic hormone, FSH, has been referred to as pleiotropic hormone in context with its multiple activities in mammals (Schulz and Miura 2002). In the testicular tissue, FSH directly stimulates its receptor (FSH-R), which is known to be expressed in Sertoli cells, promoting germ cell development and survival and thereby increasing the spermatogenetic capacity of the testicles (Hess and de Franca 2008; Rato et al. 2012b; Schulz and Miura 2002). Pharmacological and genetic experiments in rodents showed that the primary role of FSH during prepubertal development involves the stimulation of Sertoli cells proliferation, contributing to the determination of germ cells numbers in the adult individual (Heckert and Griswold 2002). Targeted mutations in the murine FSH-R led to dramatically reduced testicles weights and epididymal sperm numbers, although fertility was not lost (Dierich et al. 1998; Kumar et al. 1997). Moreover, FSH-R knockout mice and FSH knockout mice showed a significant decrease of spermatogonia, spermatocytes and spermatids numbers, suggesting that FSH may act in order to increase the number of spermatogonia and the entry of these cells into meiosis (Abel et al. 2008; Haywood et al. 2003). Indeed, administration of FSH or transgenic expression of FSH in hipogonadal mice led to an increase in the number of spermatogonia and spermatocytes and also to the formation of round spermatids (Haywood et al. 2003; Singh and Handelsman 1996). It has also been reported that a reduction in FSH levels during meiosis, particularly in the pachytene spermatocytes, caused a marked increase in germ cell death, suggesting that FSH normally acts as a germ cell apoptosis suppressor (Ruwanpura et al. 2008).

Although testosterone has no direct impact on germ cell development, it indirectly regulates spermatogenesis via Sertoli cells, being vital for the functioning of the adult testicles (Roberts and Zirkin 1991). Its testicular levels in men and rodents are 25 to 125-fold higher than those present in serum (Rato et al. 2013). As mentioned, LH modulates the regulation of testosterone synthesis. Administration of exogenous testosterone in LH-R knockout mice is able to resume spermatogenesis even in the lack of LH-R function (Lei et al. 2001). In the absence of exogenous testosterone replacement, spermatogenesis is arrested during meiosis in males lacking both LH and FSH (Iddon et al. 1977). Moreover, it has been reported that a deficiency in the production of testosterone or defects in androgen receptor results in a halt of spermatogenesis in the middle of meiosis, illustrating that testosterone action in Sertoli cells is essential for the completion of meiosis. In the absence of testosterone, there is a loss of pachytene spermatocytes and round spermatids between stages VII–VIII of spermatogenesis of mice, which can be reversed by treatment with testosterone (O'Donnell et al. 1994; O'Shaughnessy et al. 2010; Pakarainen et al. 2005; Sun et al. 1989). This loss seems to be due to a failure of Sertoli cells in producing the adhesion molecule N-cadherin, the production of which appears to require both FSH and testosterone (Perryman et al. 1996). Adhesion between Sertoli cells and spermatids is androgen dependent, since androgen action is required to prevent the retention and phagocytosis of mature, elongated spermatids and the premature release of round spermatids (O'Donnell et al. 2011; Kerr et al. 1993). Testosterone also appears to be required for spermiogenesis and spermiation (Sofikitis et al. 2008). On the other hand, there is no evidence suggesting that spermatogonial differentiation and self-renewal requires the presence of testosterone. In fact, testosterone seems to have an inhibitory action on spermatogonial differentiation in irradiated rats (Roberts and Zirkin 1991). Reduction of the testosterone action by antagonists of the androgen receptor successfully induced spermatogonial differentiation (Roberts and Zirkin 1991). Hence, it seems that testosterone has a dual action in spermatogenesis. Although the detrimental role of testosterone on spermatogenesis can only be observed either under pathological conditions or in mutated rodents, it is noticeable that the beneficial effects of testosterone are exhibited in mid-meiosis and later stages of spermatogenesis, while the detrimental effects reside in differentiating spermatogonia. One may speculate that fine-tuning of the local action testosterone during the seminiferous epithelial cycle in normal testicle is essential for a successful spermatogenesis.

3.2.2 Other Factors Involved in the Regulation of Spermatogenesis

Temperature is a key factor in regulating the progression of spermatogenesis. In many mammalian species successful spermatogenesis requires a temperature below 37 °C. In fact, male rodent gonads exposed to temperatures above 35 °C showed rapid weight loss and histological damages which varied in extent, with a reduction or absence of pachytene spermatocytes and round spermatids, and abnormally shaped spermatids (Meistrich et al. 1973). It is not clear why a temperature below 37 °C is required for the normal completion of meiosis in the spermatogenic event, but individuals with cryptorchidism (the condition where testicles fail to descend into the scrotum and reside in the abdominal cavity, at the normal body temperature) exhibit an arrest in the spermatogenic event (Yin et al.

1997). Pachytene spermatocytes seem to be the germ cell population most sensitive to hyperthermia, while spermatogonia seem the most resistant (Meistrich et al. 1973). The processes involved in spermiogenesis are also suggested to be adversely affected by an increase of the temperature (Sailer et al. 1997).

Reactive oxygen species (ROS) and testicular antioxidant defenses also influence the development of the cyclic waves of spermatogenesis. The maintenance of a high redox potential is a requirement for a healthy reproductive system (Fujii et al. 2005). If on one hand ROS are needed for the maintenance of male reproductive function, on the other hand the risk caused by ROS must be minimized using antioxidant systems. In order to address this risk, the testicular tissue has developed an array of antioxidant systems comprising of both free radical scavengers and antioxidant enzymes, such as superoxide dismutase and glutathione peroxidase. Damage occurs when the levels of ROS exceed the scavenging capacity of the redox system. In this context, cells can repair oxidized molecules using NADPH as an original electron source, being that selenium (as in the glutathione system) plays a key role (Guerriero et al. 2014). It has long been known that selenium is required for the synthesis of testosterone and the formation and development of the spermatozoa. Its deficiency affects testicular mass, with particular damages to sperm morphology and motility (Olson et al. 2005). Nevertheless, the production of ROS in mammalian male germ cells is a vital physiological event for the functional maturation and capacitation of spermatozoa (Aitken and Fisher 1994). Glutathione peroxidase catalyzes the oxidation of the bulk of protein sulfhydryl groups that takes place during the final phases of male germ cell maturation. This process promotes the assembly of the midpiece of spermatozoa and chromatin condensation, highlighting a mechanism of coupling a physiologically controlled oxidative stress to a specialized phenotypic function (Maiorino and Ursini 2002).

Chapter 4
Sertoli Cell and Germ Cell Differentiation

Germ cells may be defined as the only type of cell with the capability of generating an entirely new organism, transmitting the genetic information from one generation to the next (McLaren 2003). In male mammals, the successful progression of germ cells into fully competent spermatozoa is an extremely organized and precisely timed process that is finely controlled by Sertoli cells (Griswold 1998). Before entering puberty, the immature Sertoli cells have the ability to stimulate spermatogonia proliferation and self-renewal. Then, as Sertoli cells maturation occurs during puberty, these cells acquire new functions that make them the primary support of spermatogenesis (Griswold 1998, 1995). The establishment of Sertoli cell adult population is crucial to the formation of germ cells niches that allow a certain number of germ cells to reside in or repopulate the seminiferous tubules (Griswold 1998; Meachem et al. 2001).

Despite several studies illustrating that Sertoli cells numbers can be augmented in adults under certain conditions, those numbers are actually considered to be numerically stable in adult male mammals. In fact, it has been reported that Sertoli cells numbers may be increased in rats of more than 15–20 days of age (when rapid Sertoli cells proliferation typically stops) by inducing a delayed growth in testicular transplants or through a temporary removal of hormonal stimulation in transplanted testicles (Johnson 1986a, b). On the other hand, experiments with animal models, in which the number of Sertoli cells is altered by manipulating the testicle size and/or the spermatogenic output, revealed a relatively constant ratio between Sertoli cells and spermatids before and after the manipulation (Hess et al. 1993; Orth et al. 1988; Simorangkir et al. 1995). So, it is accepted that the number of Sertoli cells determines the number of germ cells that can be supported through spermatogenesis (Sharpe et al. 2003; Orth et al. 1988). Nevertheless this capacity varies between species. For instance, in adult rat testicles the Sertoli-germ cell ratio is approximately 1:50, while in humans the ratio of Sertoli cells to germ cells is as low as 1:11 (Hikim et al. 1985; Weber et al. 1983; Wong and Russell 1983).

© The Author(s) 2015   25
P.F. Oliveira and M.G. Alves, *Sertoli Cell Metabolism and Spermatogenesis*,
SpringerBriefs in Cell Biology, DOI 10.1007/978-3-319-19791-3_4

Within normal conditions, the number of Sertoli cells is set around puberty (for rodents around 2 weeks after birth (Orth 1984; Orth et al. 1988) and for humans at 11–13 years (Zivkovic and Hadziselimovic 2009)). However, prior to puberty, there is an early wave of spermatogonial apoptosis (resulting in 70 % of spermatogonial death), which is required to maintain the numbers of germ cells within the Sertoli cells supportive capabilities (Orth et al. 1988; Rodriguez et al. 1997). The number of Sertoli cells is also linked to the level of spermatogenesis and hence to the daily production of sperm per testicle, a factor that will obviously be reflected on the male fertility capacity (Johnson 1986b; Johnson et al. 1984; Johnson and Thompson 1983).

Overall, the role of Sertoli cells in the differentiation of germ cells has been intensively investigated and a proper functioning of these cells is known to be crucial to the preservation of male fertility. The mature Sertoli cells must create an unique microenvironment within the seminiferous tubules of the testicle, establishing the blood-testis barrier and providing the structural, nutritional and immunological support to the developing germ cells (Russell 1993b; Vogl et al. 2000). They must also produce a variety of growth factors, cytokines, signaling molecules, bioactive peptides, and several glycoproteins that form the molecular basis for Sertoli-germ cell interactions (Skinner 1993a; Griswold 1988). In fact, spermatogonial stem cell self-renewal and differentiation is dependent on the action of a range of autocrine and paracrine factors from the somatic environment, many of which are produced and delivered by Sertoli cells.

4.1 Blood-Testis-Barrier Functions

Sertoli cells are responsible for the formation of the blood-testis barrier, which facilitates the movement of the developing germ cells and the release of mature germ cells (De Kretser and Kerr 1988; Mruk and Cheng 2004). The first evidence of the existence of a "barrier" within the testicle was demonstrated in the early 1900s by the exclusion of dyes from the testicular tissue (Bouffard 1906; Ribbert 1904). Nevertheless, its existence was largely ignored for years until other studies confirmed the original observations, illustrating that certain dyes or radiolabelled material did not readily pass into the seminiferous tubules (Chiquoine 1964). Specialized junctions between adjacent Sertoli cells, located near the basement membrane, create the blood-testis barrier. These junctional complexes are located in the basal third of the seminiferous tubules (Wong and Cheng 2005). The blood-testis barrier is responsible for the division of the seminiferous epithelium into two compartments: the basal compartment (containing spermatogonia, preleptotene and leptotene spermatocytes) and adluminal (or apical) compartment (containing all the subsequent stages of spermatocytes, as well as spermatids and spermatozoa) (Mruk and Cheng 2004; Su et al. 2011). So, it serves as an anatomical/physical barrier to restrict the entry of molecules into the adluminal compartment. Moreover, this barrier confers polarity to the Sertoli cells and regulates the paracellular diffusion of water, electrolytes, nutrients and biomolecules from the systemic circulation to germ cells (Papaioannou et al. 2009), creating an enabling microenvironment for germ cells development. Any dysfunction of this barrier is expected to lead to the arrest of germ cells development and to the degeneration of germ cells (Levy et al. 1999). While the anatomical barrier restricts the passage of molecules, specific transporters located along the basolateral and apical membranes of the Sertoli cells regulate the movement of molecules in and out of the lumen of the seminiferous tubules (Setchell 2009), creating a highly dynamic physiological barrier. As the Sertoli cell changes its three-dimensional structure during the course of spermatogenesis, this physiological barrier also changes to meet the needs of the developing germ cells (Setchell 1980).

It is now accepted that mature Sertoli cells acquire a characteristic spatial arrangement that provides them the unique capability to morphologically and/ or biochemically interact with the different generations of germ cells, integrating and modulating the various signals to, from and between germ cells. Each Sertoli cell is simultaneously in contact with three or four layers of germ cells. Its base is in contact with the basement membrane and with the more undifferentiated spermatogonia. The lateral surfaces of Sertoli cells project cellular processes that enclose spermatocytes and early spermatids and its apical portion is connected to elongating and elongated spermatids (Jégou 1992; Li et al. 2009a). The disruption of the cell adhesion complex at the elongated spermatid-Sertoli cells interface only occurs during spermatozoa detachment from the seminiferous epithelium, allowing its sequential entrance to the tubular lumen for further maturation in the epididymis (Cheng et al. 2010). This well-timed movement of developing germ cells across the blood-testis barrier consists of intermittent phases of junction disassembly and reassembly, which allow the translocation and morphological transformation of germ cells during their development, while maintaining the integrity of the barrier (Chung et al. 1999). If the passage across the blood-testis barrier is accelerated, this will induce a premature detachment of germ cells from the epithelium and the produced spermatozoa are unable to fertilize the egg due to their immaturity. Likewise, if the process is hampered and germ cells become retained in the epithelium, they will be removed by Sertoli cells via phagocytosis. In either case, infertility will occur (Chung et al. 1999), illustrating the importance of a proper function in Sertoli cell.

The blood-testis barrier also acts as an immunological barrier that limits the movement of immune cells and regulates the level of cytokines in the seminiferous epithelium (Mital et al. 2011; Sikka and Wang 2008). Since the immune system capacity to distinguish between self and non-self antigens is acquired way before the first appearance of spermatocytes and spermatids at puberty, if there was no blood-testis barrier these cells would be recognized as foreign cells by the immune system and would be eliminated (Johnson et al. 2008). In fact, the immune-privilege in the testicle extends beyond the adluminal compartment of the seminiferous tubules, encompassing the whole testicle, since, although most of the autoantigenic germ cells are sequestered beyond the blood-testis barrier, preleptotene spermatocytes and spermatogonia are located within the basal compartment of the seminiferous epithelium, and express antigens that can evoke an immune response (Mital et al. 2011). The establishment of a testicular immune-privileged environment involves the local production of anti-inflammatory cytokines and immuno-regulatory factors, not only by Sertoli cells, but also by Leydig cells and testicular immune cells, that function to form a functional immunological barrier (Meinhardt and Hedger 2011). Hence, the immunological component of the blood-testis barrier combined with the other immuno-modulatory properties of the testicles, lead to the creation of an effective immune-privileged environment that, under normal circumstances, protects the autoantigenic germ cells from immunological destruction.

4.2 Sertoli-Sertoli Cell Interface Junctions

Various intercellular junctions occur at Sertoli-Sertoli cell contact sites allowing the establishment of the blood-testis barrier and maintaining epithelial cell polarity and integrity. The several types of junctions that create the blood-testis barrier structure are highly organized to guarantee a selective passage from the outside to the inside of the tubule (Mruk and Cheng 2004; Walker and Cheng 2005).

Concerning the molecular architecture of the blood-testis barrier, as it happens in other epithelia, it is known that it includes three major classes of proteins in its composition: integral membrane proteins, peripheral adaptors (and their associated signaling molecules) and cytoskeletal proteins (Wong and Cheng 2005). These constituents allow the establishment of three different types of intercellular junctions between Sertoli cells: occluding junctions (tight junctions), anchoring junctions (consisting of adherens junctions, focal contacts and desmosomes) and communicating junctions (gap junctions). The blood-testis barrier is predominantly constituted by tight junctions and basal ectoplasmic specializations. Other components of this barrier include the desmosome-like junctions, which provide adhesion that can withstand to mechanical force, and gap junctions that mediate intracellular communication (Papaioannou et al. 2009).

Within the testicle, tight junctions are only found in the blood-testis barrier (Mruk and Cheng 2004). At the beginning of meiosis, germ cells located outside the barrier must pass through the tight junctions and once beyond the blood-testis barrier they became dependent on the support exerted by Sertoli cells. Tight junctions are occluding junctions that form an area of close contact between plasma membranes of neighboring Sertoli cells, dividing these cells in a basal and an adluminal domain. Sertoli cell tight junctions, contrastingly with those found in other epithelial cells, have a distinctive localization in the plasma membrane, being placed closest to the basement membrane, completely restricting even the passage of small tracers (ferritin, peroxidase and lanthanum) into the lumen of the seminiferous epithelium (Dym and Fawcett 1970; Mruk and Cheng 2004). Hence, testicular tight junctions have two main functions: (1) establishment of the blood-testis barrier, which has been described as one of the tightest blood tissue barriers, and (2) formation of a boundary that confers polarity to the seminiferous epithelia preventing the intermixing of molecules between the two domains (basal and adluminal). Testicular tight junctional complexes are constituted by the association of integral and peripheral proteins with elements of the cytoskeleton. Only a few classes of integral membrane proteins have been described in testicular tight junctions (occludins, claudins, junctional adhesion molecules (JAMs), crumbs homolog 1 (CRB1) and coxsackievirus and adenovirus receptor (CAR)), while several types of peripheral membrane proteins have been identified at the site of testicular tight junctions (zonula occludens (ZO)-1, 2 and 3, cingulin, symplekin, among others) (Fig. 4.1) (Furuse et al. 1993; Gliki et al. 2004; Moroi et al. 1998; Mruk and Cheng 2004). In addition, multiple proteins that are known to regulate the dynamics of tight junction (by recruiting tight junction components) are also associated to these junctional complexes. Examples are signaling proteins, such as the protein kinases A and C, and vesicular transport proteins, such as the phosphatase and tensin homolog (PTEN) (Mruk and Cheng 2004).

Two modified adherens junction types are present at the interfaces between Sertoli cells, namely the ectoplasmic specialization and tubulobulbar complex. The ectoplasmic specialization is a testicle-specific atypical adherens junction characterized by the hexagonally packed actin filament bundles inserted between the endoplasmic reticulum and the plasma membrane of Sertoli cell (Grove and Vogl 1989; Russell 1977b; Vogl and Soucy 1985). When connecting Sertoli cells, the ectoplasmic specialization is present side-by-side with tight junctions and named as basal ectoplasmic specialization. The basal ectoplasmic specialization is also a component of the blood-testis barrier and is composed of at least two types of adherens junction transmembrane proteins [classic cadherins (Lee et al. 2003; Wine and Chapin 1999) and nectin-2 (Mueller et al. 2003; Ozaki-Kuroda et al. 2002)], cross-linking directly or indirectly with the bundles of actin filaments through their binding proteins (afadin and catenins, respectively) (Fig. 4.1) (Wong et al. 2008). The tubulobulbar complex is another testicle-specific adherens junction that can be found between Sertoli cells at the same level of tight junctions (Vogl et al. 2000). The appearance of these junctions is associated and precedes two fundamental events during spermatogenesis: spermatocyte translocation from basal to adluminal compartment and sperm release from the seminiferous epithelium (Vogl et al. 2012). Like ectoplasmic specializations, the tubulobulbar complexes are present at two major locations in the seminiferous epithelium: at apical sites, establishing intercellular connections between Sertoli cells and developing germ cells, and at basal sites, promoting the attachment of neighboring Sertoli cells (Upadhyay et al. 2012). In general, a tubulobulbar complex consists of a tubular process of one Sertoli cell that extends into a correspondent invagination on a neighboring Sertoli cell. Each tubulobulbar complex consists of: (1) a long proximal tubule, (2) a bulb, (3) a distal tubule, and (4) a clathrin-coated pit that covers the terminal end of distal tubule. Actin-filament networks, intimately connected with the two tubular structures, and cisternae of endoplasmic reticulum, which are associated with the bulb, are also other components of this complex (Upadhyay et al. 2012). Other important components have been identified in the

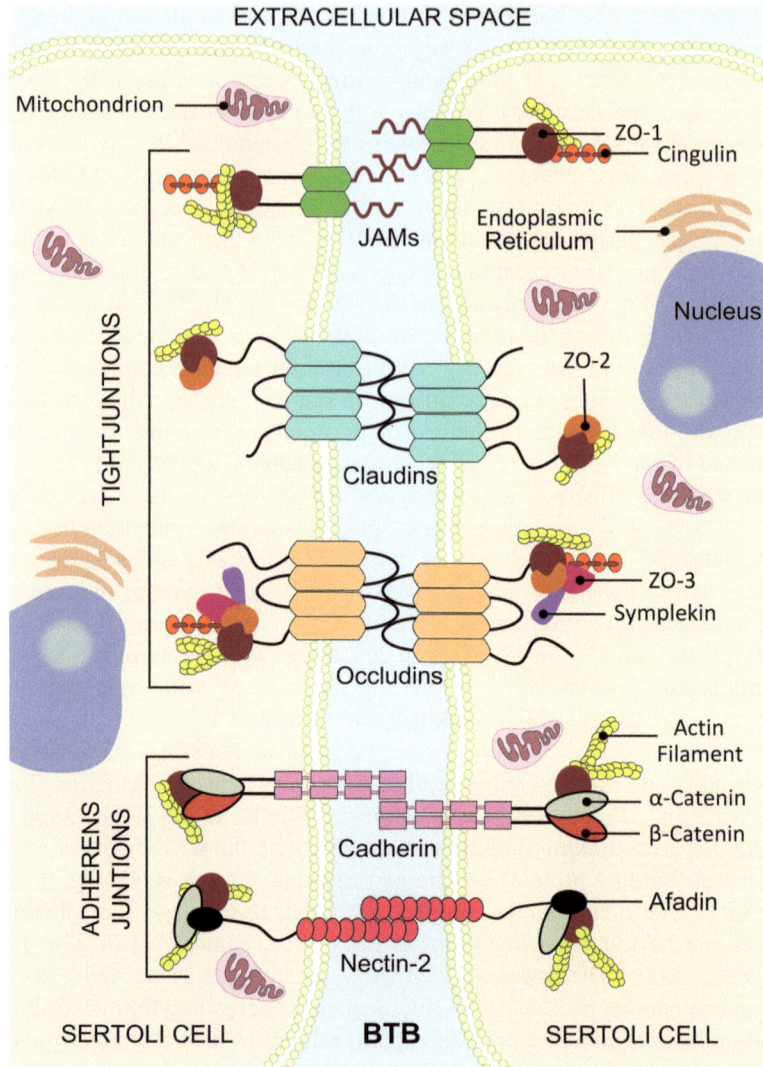

Fig. 4.1 Schematic representation of the main transmembrane junctional complexes existing between two adjacent Sertoli cells, which are commonly known as blood-testis-barrier (*BTB*). This barrier is formed by tight and adherens junctions that coexist side-by-side to preserve blood-testis-barrier integrity. Junctional adhesion molecules (*JAMs*), occludins ans claudins are the major proteins of blood-testis-barrier tight junctions. They are linked to the actin-based cytoskeleton via several regulatory proteins, such as zonula occludens (ZO-1,-2 ans -3), cingulin and symplekin, which are essential to enhance cellular adhesion between two neighboring Sertoli cells. The adherens junctions between Sertoli cells include cadherin and nectin-2, which directly interact with actin-based cytoskeleton through their binding proteins, α/β-catenin and afadin, respectively. Abbreviations: *BTB*—Blood-testis barrier; *JAMs*—Junctional adhesion molecules; *ZO*—zonula occludens

tubulobulbar complex, including dynamins and amphiphysins (Upadhyay et al. 2012). It is suggested that basal tubulobulbar complex may assist the internalization of other types of junctions during germ cell movement (Vogl et al. 2012). Yet, due to the difficulties in identifying and observing these basal complexes, few information is available on their specific structure and significance.

Desmosomes are another type of cell-cell adherens junctions present between adjacent Sertoli cells. These anchoring junctions are constituted by two main domains: (1) an extracellular core or desmoglea and (2) cytoplasmic plaques that lie parallel to the plasma membrane (Mruk and Cheng 2004). The symmetrical dense cytoplasmic plaques use intermediate filaments as their attachment sites, forming a continuous network throughout the entire cell. The Sertoli cell desmosomes are constituted by several of the same proteins that constitute desmosomes in other tissues, namely desmocollins, desmogleins, desmoplakins, plakophilins and plakoglobins (Mruk and Cheng 2011). Sertoli-Sertoli cell desmosomes are juxtaposed with tight junctions and basal ectoplasmic specializations, functioning in cell-cell adhesion. Additionally, since multiple protein kinases have been reported to be associated with desmosomal proteins, it is accepted that these junctional complexes also provide important platforms for signal transduction events that control various aspects of the cell function (Mruk and Cheng 2011).

Gap junctions are the last type of intercellular junctional complexes that have been described between Sertoli cells. These communicating junctions are known to mediate the exchange of ions and small molecules between two adjacent cells, via intercellular channels formed by the noncovalent interaction of two hemichannels (Kopera et al. 2010). Gap junctions are formed by various connexins, which have highly conserved transmembrane and extracellular domains, but divergent regions located between the transmembrane and C terminal domains. Although the presence of gap junctions between Sertoli cells has been known since the 1970s (McGinley et al. 1979), up until now very little is known about how gap junctions contribute to the physiology of the Sertoli cell and the seminiferous epithelium.

4.3 Sertoli-Germ Cell Interface Junctions

Sertoli-germ cell adhesion is also a key process in the spermatogenic event. The disruption of the Sertoli–germ cell association leads to a breakdown of spermatogenesis. Factors such as heat, cytotoxic agents and several pathological conditions are known to contribute for the occurrence of the disruption of this association (Hecht 2002). Sertoli-germ cell adhesion is maintained by cell-cell actin-based adherens junctions and intermediate filament-based desmosome-like junctions, being that the status of these junction complexes is dependent of the developmental stage of germ cells (Cheng and Mruk 2002). For instance, desmosome-like junctions are formed between primary spermatocytes or round spermatids and Sertoli cells (Mruk and Cheng 2004; Wong et al. 2008). When round spermatids

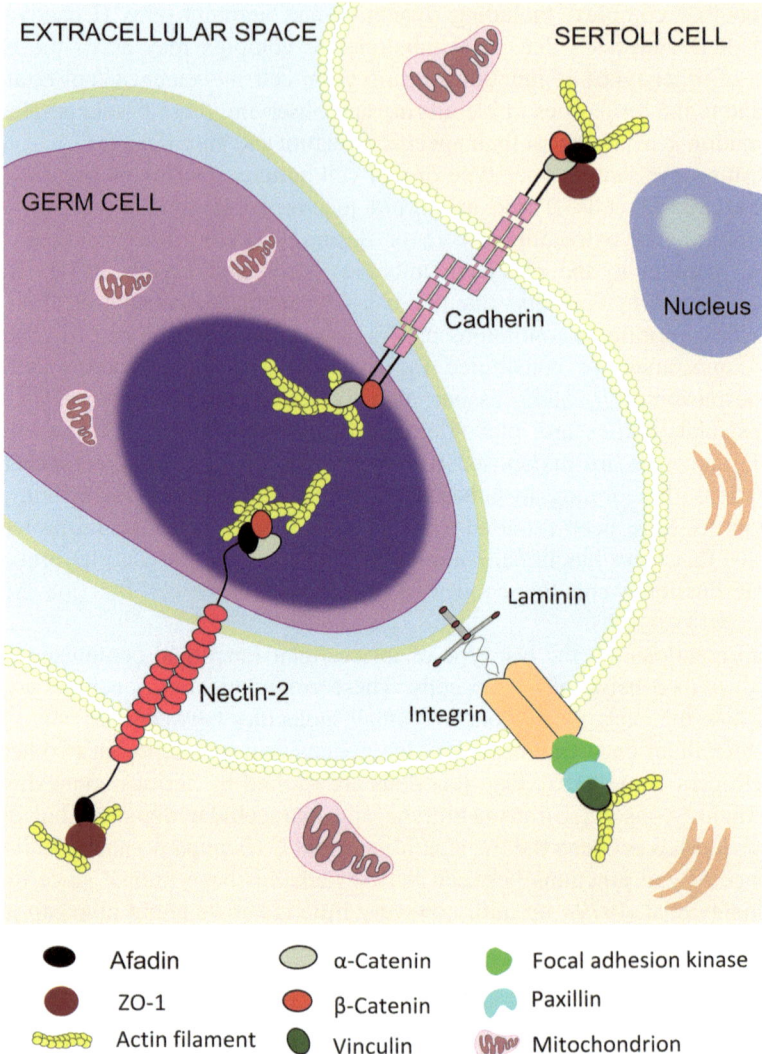

Fig. 4.2 Schematic representation of the molecular architecture of the Sertoli-germ cell interface junctions illustrating Cadherin/Cadherin, Laminin/Integrin and Nectin-2/Nectin-2 junctions. Cadherin in the germ cell surface bind to catenin members, α and β, while cadherin in the Sertoli cell surface, is not only linked to catenins but also to the adhesion molecule zona occludens-1 (ZO-1) and afadin protein. Within germ cells, nectin-2 is associated with afadin, α-catenin and β-catenin, whereas nectin-2 on Sertoli cells binds to afadin, α-catenin and ZO-1. The laminin present on germ cells surface interacts with Sertoli cells integrin complexes activating the focal adhesion kinase, which in turn associates and activates paxillin and vinculin proteins. All the junctions are directly linked with actin filaments promoting a firm cell adhesion between Sertoli cells and germ cells. Abbreviations: *ZO*—zonula occludens

start to elongate, this anchoring junction is replaced by the apical ectoplasmic specializations (Russell 1977b; Toyama et al. 2003). While the ultrastructure of the actin-based cell–matrix junction is absent in the testicle, components of the focal contact were found at the apical ectoplasmic specializations, including integrin, laminin, focal adhesion kinase, paxillin, and vinculin (Fig. 4.2) (Mruk et al. 2008; Yan and Cheng 2006). The apical ectoplasmic specialization provides mechanical adhesion of spermatids onto the nourishing Sertoli cells, facilitating the movement of developing spermatids across the epithelium and ensuring the proper orientation of spermatids so that fully developed spermatids can be released to the tubule lumen during spermiation. Without this controlled event of spermatid movement, spermatogenesis cannot be completed (Siu et al. 2005). The tubulobulbar complex is another junctional complex that can be observed on the concave side of elongated spermatids a few hours before spermiation and it is mutually exclusive with the apical ectoplasmic specialization. It extends from the concave surface of the spermatid head towards the Sertoli cell. A unique characteristic of apical tubulobulbar complexes is that they are not visible in the epithelium until a few days before spermiation at stage VII (Russell and Clermont 1976; Russell 1979b).

Communicating gap junctions are also established between Sertoli and germ cells. They are similar in structure and composition as those established between adjacent Sertoli cells (McGinley et al. 1979). Although, as previously referred, few is known on the contribution of gap junctions to the physiology of the seminiferous epithelium and particularly of spermatogenesis, indirect observations with connexin–null male mice have shown that they likely play a crucial role in coordinating events pertinent to germ cell movement in the seminiferous epithelium (Juneja et al. 1999). As gap junctions mediate signals between Sertoli and germ cells, they likely ensure Sertoli-germ cell metabolic and signaling coupling, allowing the synchronization of male germ cell proliferation and differentiation.

4.4   Sertoli Cells Secreted Factors Influencing Germ Cells Development

Sertoli cells secrete a wide variety of factors required for the successful development of germ cells into spermatozoa (Table 4.1). For instance, Sertoli cells dedicate a reasonably large percentage of their protein synthesis to the production and secretion of glycoproteins necessary for the developing germ cells (Griswold 1995). Likewise, there is also a substantial transport of germ cells molecules that affect Sertoli cells function and cyclicity (Fujisawa et al. 1992; Pineau et al. 1990; Stallard and Griswold 1990), supporting the existence of a bidirectional interaction between these two types of cells.

Among the numerous substances secreted by Sertoli cells, the key factors required for a successful spermatogenesis can be divided into categories according to their known biochemical properties: (1) those required to fuel germ cells differentiation and metabolism (Skinner 2005; Sylvester 1993); (2) glycoproteins that

Table 4.1 Main factors produced by Sertoli cells involved in the regulation of germ cell proliferation, survival and differentiation

Factor	Putative effect	References
Cell differentiation and metabolism		
Lactate, acetate and pyruvate	Germ cells energy supplies	Boussouar and Benahmed (2004)
Activin and Inhibin	Stimulation/inhibition of spermatogonial proliferation	Sikka and Wang (2008)
TGF-β	Germ cells division and differentiation	Jégou (1993)
SCF, FGF2, BDNF and LIF	Spermatogonial stem cells proliferation and survival	Kadam et al. (2013), Kanatsu-Shinohara and Shinohara (2013), Zhang et al. (2012), Ebata et al. (2011), Allard et al. (1996)
GDNF	Stimulates replication, suppresses differentiation and induces migration of spermatogonial stem cells	Meng et al. (2000), Sikka and Wang (2008)
BMP4 and SCF	Spermatogonia stem cell differentiation	Hu et al. (2004), Pellegrini et al. (2003), Van Pelt and de Rooij (1990)
IGF1	Stimulates the last steps of spermatogenesis	Itoh et al. (1994)
Interleukin-1α	Influences transferrin expression and lactate production in these cells	Huleihel and Lunenfeld (2002), Nehar et al. (1998)
Interleukin 6	Inhibits meiotic DNA synthesis during the spermatogenic cycle and influences the secretion of transferrin and inhibin B	Boockfor and Schwarz (1991)
Transport binding proteins		
Transferrin and Ceruloplasmin	Iron and copper transport into germ cells	Le Magueresse et al. (1988) and Sylvester (1993)
ABP	Increase in concentration of sex steroid hormones within the luminal fluid of the seminiferous tubules	Sylvester (1993)
Sertoli-to-germ cell signalling		
SCF	Interaction with c-kit transmembrane tyrosine kinase receptor present on spermatogonia Mediate cellular adhesions between Sertoli cells and spermatocytes	Sofikitis et al. (2008), Manova et al. (1990), Yoshinaga et al. (1991), Eddy (2002)
LIF and IGF2	Interaction with IFG2/M6P receptor	Tsuruta and O'Brien (1995), O'Brien et al. (1993)
FGF4	Involved in Sertoli–spermatogenic cell interaction	Yamamoto et al. (2002)

(continued)

Table 4.1   (continued)

Factor	Putative effect	References
Blood-testis barrier dynamics		
Proteases and proteases inhibitors (Cathepsins, plasminogen activators)	Involved in blood-testis barrier assembly/disassembly and in basement membrane remodeling	Wong and Cheng (2005)
TGF-α, -β and -β3	Regulation of the levels of tight junction and adherens junction integral membrane proteins, proteases, protease inhibitors and extracellular matrix proteins in the seminiferous epithelium	Siu et al. (2003), Lui et al. (2001)
Interleukin 6	Modulation of the blood-testis barrier dynamics	Zhang et al. (2014)
Basement membrane components		
Collagen types I and IV, Laminin, Proteoglycans and fibulins	Required for the tubular cyto-architecture, for cell to cell communication and for the storage and action of various factors	Pineau et al. (1999)

function as growth factors or paracrine factors (Sikka and Wang 2008; Skinner 2005); (3) proteases and proteases inhibitors that have a role in tissue remodeling processes occurring, for instance, during spermiation or movement of preleptotene spermatocytes into the adluminal compartment of the seminiferous tubule (Fritz et al. 1993) (4) hormones that, in addition to the secretion of growth factors, play important regulatory roles and (5) structural components of the basement membrane between the Sertoli cells and the peritubular cells (Skinner 1993b).

Pyruvate, acetate and lactate are examples of metabolic precursors produced by Sertoli cells that are required for germ cell metabolism (Rato et al. 2012b). As will be intensively discussed in the subsequent chapters, these metabolites are important not only for ATP production, but also for the differentiation and survival of the developing germ cells that are unable to metabolize other substrates (Boussouar and Benahmed 2004). Besides producing large amounts of metabolic precursors, Sertoli cells also secret a large quantity of glycoproteins (Griswold 1995). Transferrin, ceruloplasmin and the androgen binding protein (ABP) are amongst the most abundant products of Sertoli cells, having distinct functions. While transferrin and ceruloplasmin have an important role in iron and copper transport into germ cells, respectively, ABP is involved in the increase of sex steroid hormones concentration within the luminal fluid of the seminiferous tubules (Sylvester 1993; Le Magueresse et al. 1988).

Several other regulatory glycoproteins produced by Sertoli cells are secreted in much lower abundance, carrying out their biochemical roles as growth factors (Griswold 1995). The transforming growth factors alpha and beta (TGF-α and TGF-β) are Sertoli cell secreted glycoproteins distinctively important for the dynamics of the blood-testis barrier, due to their effects on the expression levels of tight junctions and adherens junctions integral membrane proteins, and also on the levels of proteases, protease inhibitors, and extracellular matrix proteins in the seminiferous epithelium (Watrin et al. 1991; Wong and Cheng 2005). For instance, the amount of TGF-α, produced by Sertoli cells, significantly declines during the assembly of Sertoli cell tight junction barrier in vitro, illustrating that this growth factor can modulate the function of blood-testis barrier (Siu et al. 2003). Furthermore, the expression of TGF-β3, which is the most abundant isoform of TGF-β in the testicle, is the highest at the onset of puberty in rats (Mullaney and Skinner 1993) and in vitro its level declines when Sertoli cell tight junction barrier is being assembled (Lui et al. 2001). TGF-β also seems to be involved in germ cell division and differentiation, as their specific receptors are differentially expressed on developing germ cells (Jégou 1993).

Sertoli cells promote the differentiation of spermatogonial stem cells through the action of the bone morphogenetic protein 4 (BMP4). BMP4 is a member of the TGF-β superfamily, secreted by Sertoli cells from a very early period in the postnatal life, with its levels being gradually downregulated until puberty. In adult testicles, BMP4 is also secreted by spermatogonia, implying that BMP4 may act via paracrine and autocrine pathways for spermatogonia stem cell differentiation during the different stages of seminiferous epithelial development (Hu et al. 2004; Pellegrini et al. 2003).

Several insulin-like growth factors (IGFs) with different functions are also produced by Sertoli cells. IGF1 specifically stimulates the last steps of spermatogenesis (Itoh et al. 1994), while IGF2 has been implicated in Sertoli-to-germ cell signaling (Eddy 2002). In fact, the bi-functional IGF2 receptor (also called the cation-independent mannose 6-phosphate receptor or M6P-R) is an important collaborator on Sertoli-to-germ-cell signaling. This receptor is known to be present on the germ cells surface and several ligands for this receptor are secreted by Sertoli cells, namely IGF2 or other M6P-bearing glycoprotein ligands that remain to be characterized (O'Brien et al. 1993; Tsuruta and O'Brien 1995).

The family of fibroblast growth factors (FGFs) encompasses numerous members, which are key players in the processes of proliferation and differentiation of a wide variety of cells and tissues. Several FGFs are secreted by Sertoli cells and have been shown to influence testicular function. The basic FGF (also known as bFGF or FGF2) is one of the most studied FGF family members and its expression has been described in Sertoli cells. Recently, it has been demonstrated that the FGF2 is essential for the proliferation and maintenance of spermatogonial stem cells (Ebata et al. 2011; Kanatsu-Shinohara and Shinohara 2013; Zhang et al. 2012). FGF4 is also expressed by Sertoli cells and its silencing results in impaired fertility, while a knock-in resulted in enhanced spermatogenesis (Yamamoto et al. 2002). These observations illustrate that the expression of FGF4 by Sertoli cell is important for normal testicular function, showing a potential for this FGF on Sertoli cell–spermatogenic cell interaction (Yamamoto et al. 2002). FGF9 is another member of the FGF family expressed by Sertoli cells, which has been shown to be essential for the early development of embryonic testicle. FGF9 is known to influence male sex differentiation, with its knockout resulting in sex reversal and impaired development of the testicles (Colvin et al. 2001).

Leukemia inhibitory factor (LIF), and the precursor forms of TGF-β are other M6P-glycoprotein-containing growth factors found in the testicle that may act through the IGF2 receptor known to be present on germ cells. IGF2 receptors are abundant in the cells of the seminiferous epithelium. While Sertoli cells predominantly express the IGF2 receptor on intracellular membranes, germ cells express their IGF2 receptors primarily on the cell surface (De Miguel et al. 1996). Several studies showed that interactions between M6P-containing ligands and germ cells can occur in both, the basal and luminal compartments, of the seminiferous epithelium, with the possibility that different ligands are secreted into each compartment (Tsuruta et al. 2000; Tsuruta and O'Brien 1995). In fact, during spermatogenesis, IGF2 receptors are mostly abundant in spermatogonia and early spermatocytes, with studies indicating that LIF-mediated signaling events involving both somatic and germ cells are important for spermatogenesis (Jenab and Morris 1998).

Cytokines are also among the glycoproteins expressed and/or exported by Sertoli cells. Although they are classically defined as growth factors involved in immune cell communication, cytokines also influence other cells and tissues. For instance, Sertoli cells express interleukin-1α, which directly influences transferrin expression and lactate production in these cells (Huleihel and Lunenfeld 2002; Nehar et al. 1998). The production of interleukin-1α seems to be dependent on the presence of germ cells (Huleihel and Lunenfeld 2002). Interleukin-6 is also produced by Sertoli cells in response to the autocrine action of interleukin-1α. This cytokine seems to have a role in the modulation of the blood-testis barrier dynamics (Zhang et al. 2014), inhibiting meiotic DNA synthesis during the spermatogenic cycle and influencing the secretion of transferrin and inhibin B by Sertoli cells (Boockfor and Schwarz 1991).

The stem cell factor (SCF or c-kit ligand) is another cytokine secreted by Sertoli cells that is involved in the highly controlled Sertoli-to-germ cell signaling (Sofikitis et al. 2008). The c-kit transmembrane tyrosine kinase receptor is present on spermatogonia (Manova et al. 1990; Yoshinaga et al. 1991). The interaction of SCF with its receptor has proven to stimulate spermatogonia survival and proliferation (Allard et al. 1996) and a single mutation in SCF gene leads to the disruption of male germ cells development (Besmer et al. 1993). At the beginning of spermatogenesis the production of SCF by Sertoli cells dramatically changes from the soluble form to the membrane-bound form, suggesting an important role for the SCF–c-kit system in spermatogonial differentiation (Blanchard et al. 1998). It also has been reported that Sertoli cells from mice mutant for the membrane-bound form of SCF are unable to bind spermatocytes (Marziali et al. 1993), illustrating that the c-kit-receptor and SCF also mediate cellular adhesion between Sertoli cells and spermatocytes (Eddy 2002).

Sertoli cells also secrete neurotrophic factors, which are known to play important roles in cell survival, differentiation and migration (Skaper 2012). The brain-derived neurotrophic factor (BDNF) is a glycoprotein known to promote cell survival and is known to be secreted by Sertoli cells. Nevertheless, there is little information regarding the functional roles of BDNF in the male reproductive tract and particularly on spermatogenesis. Contrastingly, the glial cell line-derived neurotrophic factor (GDNF), a member of the TGF-β superfamily, is a major growth factor produced by Sertoli cells that has the ability to stimulate replication, suppress differentiation and induce migration of spermatogonial stem cells, acting via the ret/GFR receptor (Meng et al. 2000).

While Sertoli cells respond to hormones, as for instance FSH, they are also hormone-producing cells. The hormones produced by Sertoli cells play important regulatory roles on the outcome of spermatogenesis. The two closely related activin and inhibin protein hormones are products of inhibin gene. The two α and β subunits form a dimer and produce inhibin, whereas a β homodimer produces activin. Although the major role of activin and inhibin is to regulate FSH production by activation or feedback inhibition of the pituitary, several local actions for these hormones have been described within the testicles. In fact, activin and inhibin have the ability to either stimulate or inhibit both Sertoli cell and spermatogonial proliferation, thus regulating the production of sperm (Sikka and Wang 2008). The local actions mediated by inhibin and activin on the testicular environment become difficult to untangle due to the fact that they are produced not only by Sertoli cells, but also by other testicular cell types. Leydig cells and peritubular cells can secrete both inhibin and/or activin (Mather et al. 1990), which then may modulate Sertoli cell function. Sertoli cells respond to activin that can influence the proliferation and function of postnatal Sertoli cell (Buzzard et al. 2003). Inhibin and activin also modulate germ cells development and germ cells can influence the expression of inhibin by Sertoli cells (Mather et al. 1990; Mather et al. 1997).

The homeostasis of proteases and protease inhibitors produced by Sertoli cells is of extreme importance because once it is disturbed it alters the biochemical composition of the extracellular matrix leading to the opening of the cell-matrix interface and facilitating cell movement (Wong and Cheng 2005). A large variety of proteases and protease inhibitors are produced by Sertoli cells, with some of them being predominantly, or even exclusively, expressed in the testicles (Jégou 1993). Proteases can be directly involved in the removal of junctional components during blood-testis barrier disassembly to facilitate preleptotene/leptotene spermatocyte migration, or they can indirectly regulate junction turnover by activating other biologically active molecules, such as growth factors, cytokines, and extracellular matrix components (e.g. collagens). Conversely, protease inhibitors limit the activity of proteases so that proteolysis can be confined to a specific microenvironment within the epithelium (Wong and Cheng 2005). Studies showed that several proteases and protease inhibitors are involved in the regulation of blood-testis barrier dynamics (Wong and Cheng 2005). For instance, it is known that Sertoli cells express and secrete proteases, such as: (1) cathepsins, which exhibit secreted and lysosomal forms, and play a role in the degradation of residual bodies (Anway et al. 2004); (2) plasminogen activators, which are highly specific serine proteases that convert latent plasminogen into the active protease plasmin and appear to be involved in the junctional restructuring step occurring at stages VII-VIII of the rat spermatogenic cycle (Sigillo et al. 1998), and (3) protease inhibitors, such as α_2-macroglobulin and cystatin C. α_2-macroglobulin appears to inactivate proteases released by degenerating germ cells and also limit the action of growth factors and proteases involved in junctional remodeling (Zhu et al. 1994). Cystatin C is a cysteine protease inhibitor that is suggested to participate as a protection mechanism against inappropriate activity of these proteases, particularly concerning the activity of cathepsin L (Tsuruta et al. 1993), which is involved in the processes of assembly and disassembly of Sertoli-Sertoli cell and Sertoli-germ cell junctional complexes.

Finally, several of the extracellular matrix components secreted by Sertoli cells, such as collagen type I and IV, laminin, proteoglycans and fibulins, are required for the tubular cytoarchitecture, for cell to cell communication and for the storage and action of various of the growth factors previously mentioned (Pineau et al. 1999).

Chapter 5
Testicular Metabolic Cooperation

The mechanisms responsible for controlling the production of viable sperm remain largely unknown. Although for many years the metabolic cooperation between testicular cells has been overlooked, recent advances provided compelling evidence that this is a crucial event for the occurrence of a normal spermatogenesis (Rato et al. 2012b). The Sertoli cells provide an adequate and protected environment for the developing germ cells. These processes are under a very tied control and any alteration may result in dramatic consequences for male fertility (Alves et al. 2013a). As the seminiferous epithelium is physically divided in basolateral and adluminal compartments by the blood-testis barrier, the Sertoli cells are responsible for the maintenance of different levels of substances between the *rete testis* fluid and the lymph or plasma (Setchell et al. 1969). Thus, the germ cells, which are located beyond the blood-testis barrier, are dependent on the selective passage of substances through the Sertoli cells and the production of metabolites by Sertoli cells (Wong and Cheng 2005).

The relevance of Sertoli cell to spermatogenesis is thoroughly discussed in the previous chapters. However it can be summarized in two facts: Sertoli cells number is associated with the testicle size (Sharpe et al. 2003) and each Sertoli cell can only support a limited number of germ cells (Weber et al. 1983). In addition, the Sertoli cell controls the pH and ionic composition of the seminiferous fluid (Oliveira et al. 2009a, b; Rato et al. 2010). Of note, these cells are also often referred as 'nurse cells' because they provide the nutritional support required by the developing germ cells (Oliveira et al. 2009a, b; Griswold and McLean 2006; Mita and Hall 1982). The latter function was overlooked for many years. However,

© The Author(s) 2015
P.F. Oliveira and M.G. Alves, *Sertoli Cell Metabolism and Spermatogenesis*,
SpringerBriefs in Cell Biology, DOI 10.1007/978-3-319-19791-3_5

there is a close metabolic relationship between Sertoli cells and developing germ cells. The Sertoli cells secrete not only peptides, such as prodynorphin, but also nutrients and metabolic intermediates (Griswold and McLean 2006; Griswold 1998). It is imperative that germ cells receive an adequate level of energy substrates from Sertoli cells and it has long been shown that the transference of metabolic products such as amino acids, carbohydrates, lipids, vitamins and metal ions occurs between these cells (Boussouar and Benahmed 2004; Erkkila et al. 2002; Jutte et al. 1981; Mruk and Cheng 2004; Riera et al. 2001; Riera et al. 2002; Robinson and Fritz 1981).

Sertoli cell metabolism is a crucial event for the normal occurrence of spermatogenesis and it presents some unique characteristics that deserve special attention (Oliveira et al. 2014b). It has been reported that cultured Sertoli cells convert the majority of glucose to lactate (Robinson and Fritz 1981), preferring a less effective pathway in terms of ATP production, when comparing with the Krebs cycle. In fact, in these cells, only 25 % of the pyruvate produced from glucose is oxidized via the Krebs cycle (Grootegoed et al. 1986). Thus, compelling evidence clearly shows that the Sertoli cell presents a Warburg-like metabolism and prefers the fermentative instead of oxidative metabolism of glucose, although the former is in theory less efficient (Oliveira et al. 2014b). Exogenous pyruvate is also oxidized at very low concentrations when compared with glucose (Grootegoed et al. 1986) illustrating that glucose is the most reliable substrate to Sertoli cells. Thus, similarly to what happens in cancer cells, the Sertoli cells exhibit a high glycolytic flux, though it is not accompanied with cellular proliferation (Oliveira et al. 2014b).

5.1 The Glycolytic Pathway in the Sertoli Cell

Glucose is present in extremely low levels in the seminiferous tubular fluid due to its prompt metabolism (Robinson and Fritz 1981; Voglmayr et al. 1966). However, it has long been shown that glucose is essential for spermatogenesis in vivo (Zysk et al. 1975; Mancine et al. 1960). Of note, in some mammalian species the ratio of glucose concentration in tubular fluid to that in plasma is less than 0.02, while in others the concentration of glucose can be as high as 75 % of that in plasma (Setchell and Waites 1975; Setchell 1970). This different concentration found in both compartments illustrates that glucose is readily available to the cells in the basal compartment of the seminiferous tubule, but not to germinal cells in the adluminal compartment. However, non-metabolizable sugars, such as 3-0-methyl-glucose, penetrate the seminiferous tubule barrier to enter the adluminal compartment (Middelton 1973; Middleton and Setchell 1972), illustrating that low levels of glucose in the intratubular fluid may not result from an inability of this sugar to penetrate the blood-testis barrier, but instead from its metabolism to various products by one or more cell types in the seminiferous tubule (Robinson and Fritz 1981).

Glucose is a hydrophilic and polar molecule and, as a result, it can cross the lipid bilayer by simple diffusion in a very inefficient manner. Therefore, there are specific carrier proteins that facilitate its diffusion along a concentration gradient. There are two families of glucose transporters: the sodium dependent glucose transporters (SGLTs), also known as solute carrier family 5 (SLC5) and the glucose transporters (GLUTs), also known as solute carrier family 2 (SLC2) (Joost and Thorens 2001; Scheepers et al. 2004). These families are composed of a different number of transporters. The SGLTs family is composed by six different active transporters, SGLT1 to SGLT6, while the GLUTs family is composed by 14 isoforms known as GLUT1 to GLUT14 (Scheepers et al. 2004). The GLUTs family is divided in three subfamilies: the class I, which encompasses GLUT1-4; the class II which encompasses GLUT5, 7, 9 and 11; and the class III which encompasses the GLUT6, 8, 10, 12 and the H^+-coupled myo-inositol transporters. The GLUTs are distributed in a wide variety of tissues (Joost and Thorens 2001) and are responsible for mediating passive glucose transport through membranes (Klip et al. 1994). Different GLUTs were already identified in the testicle: GLUT1 (Ulisse et al. 1992), GLUT2 (Kokk et al. 2004), GLUT3 (Burant and Davidson 1994), GLUT5 (Burant et al. 1992) and GLUT8 (Doege et al. 2000). Specifically in Sertoli cells, GLUT1, GLUT2, GLUT3 and GLUT4 (Carosa et al. 2005; Galardo et al. 2008; Kokk et al. 2004; Ulisse et al. 1992) have been identified. However, not all GLUTs are expected to contribute in the same manner to glucose transport in Sertoli cells. For instance, GLUT8 has not been localized at the plasma membrane, excluding the role of this isoform in the transport of glucose from the extracellular *milieu* (Piroli et al. 2002; Reagan et al. 2001). Moreover, recent studies using a GLUT8 knockout mice model, provided evidence that this transporter is mainly involved in the transport and recycling of glucose residues in the membranes of lysosomes (Adastra et al. 2012; Carayannopoulos et al. 2000). On the other hand, GLUT1, GLUT2 and GLUT3 have been identified in the plasma membrane of Sertoli cells and therefore it may be assumed that they are the primary responsible for glucose import in these cells. However, studies in cultured mammalian cells suggest that glucose itself may regulate its own transport, since its absence or deprivation leads to compensatory mechanisms in glucose uptake to optimize the utilization of this sugar and maintain the energy levels (Klip et al. 1994). Indeed, in Sertoli cells, it has been reported that when glucose is removed from the culture medium, an increase in GLUT1 and a decrease in GLUT3 expression levels can be observed (Riera et al. 2009), which illustrates that alterations of glucose levels in the extracellular *milieu* act as a signal for Sertoli cell to upregulate the glucose membrane transport machinery in order to ensure the appropriate glycolytic flux.

Once glucose enters the cell, it suffers a series of multi-step reactions catalyzed by various enzymes. The first rate-limiting step in glucose metabolism is mediated by phosphofructokinase (PFK) that catalyzes the irreversible conversion of fructose-6-phosphate to fructose-1,6-bis-phosphate (Chehtane and Khaled 2010). The functioning of this enzyme is associated with the energy state of the cell (Mor et al. 2011) and is expected to play a crucial regulatory role in the control

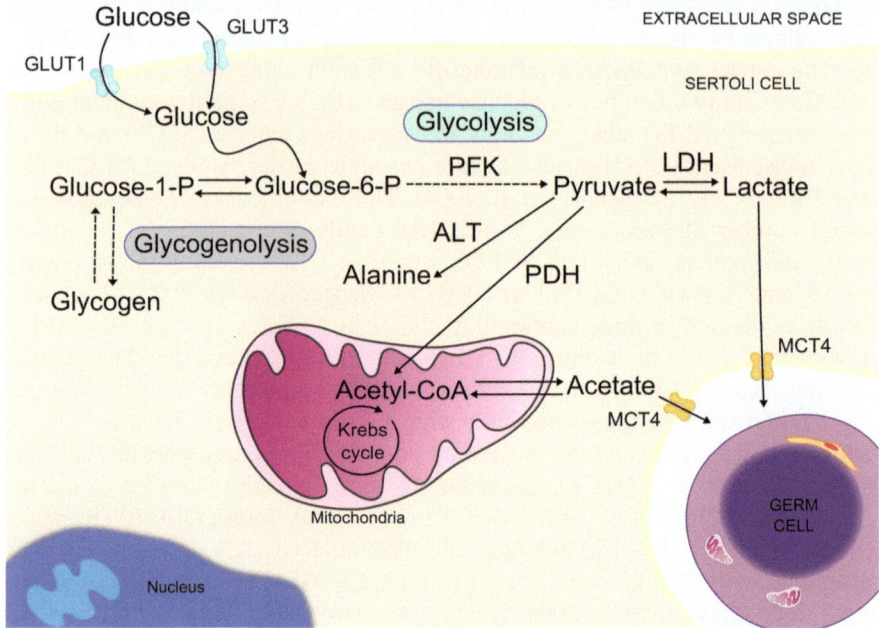

Fig. 5.1 Schematic diagram specifying the most relevant mechanisms of the metabolic cooperation established between the Sertoli cell and developing germ cells. In Sertoli cells, glucose from interstitial space is taken through high-affinity glucose transporters, *GLUT1* and *GLUT3*, which are present in the plasma membrane. When glucose is not available, glycogen is mobilized. In physiological conditions, the majority of glucose is converted to pyruvate through a rate-limiting process catalyzed by the enzyme phosphofructokinase (*PFK*), that catalyzes the irreversible conversion of fructose-6-phosphate to fructose-1,6-bis-phosphate. Pyruvate originated from glycolysis can follow three distinct paths. It can be converted to alanine by the action of alanine aminotransferase (*ALT*); it can be converted into acetyl-CoA by the action of pyruvate dehydrogenase (*PDH*); or it can be converted to lactate by the action of lactate dehydrogenase (*LDH*). Acetyl-CoA enters the mitochondria to be used in the Krebs cycle, and/or can be converted into acetate. Both acetate and lactate are exported to the intratubular fluid by specific monocarboxylate transporters (*MCT4*). These substrates are then taken up by developing germ cells. Abbreviations: GLUT1 and GLUT3 glucose transporters 1 and 3; *PFK* phosphofructokinase; *ALT* alanine aminotransferase; *LDH* lactate dehydrogenase; *PDH* pyruvate dehydrogenase; *MCT4* monocarboxylate transporter 4

of Sertoli cell metabolism (Fig. 5.1) (Martins et al. 2013, 2014). Glucose is then converted to pyruvate and the glycolytic process is completed. The cytosolic pyruvate originated from glycolysis can follow three main distinct paths. It can be converted to alanine by the action of alanine aminotransferase; it can enter the Krebs cycle; or it can be converted to lactate by the action of lactate dehydrogenase (LDH), with the simultaneous oxidation/reduction of NADH to NAD$^+$. Compelling evidence shows that the majority of the pyruvate produced by Sertoli cells is converted to lactate by LDH (Oliveira et al. 2014b).

The lactate produced by the Sertoli cells is then exported through specific monocarboxylate transporters (MCTs). The family of MCTs, also known as SLC16A, is composed of 14 members that differ in their transport kinetics and subcellular transport. From these, only MCT1 to 4 have been functionally characterized as proton-linked MCTs and proven to function on the transport of monocarboxylates (Halestrap 2012; Pellerin 2003) and thus, are thought to be involved in the transport of lactate. However, they differ in the affinity for the substrate and also on the selectively. MCT1, MCT2 and MCT4 are widely expressed in all tissues while MCT3 is specifically expressed in retina (Brauchi et al. 2005). MCT2 can be found in elongated spermatids (Goddard et al. 2003) and the presence of MCT1 and MCT4 has been reported in Sertoli cells (Galardo et al. 2007). Of note, MCT1 was identified in germ cells (Goddard et al. 2003) and it has a higher affinity and a major role in the import of lactate from the extracellular *milieu* (Bonen 2001), while MTC4, which has a lower affinity for lactate, is primarily a lactate exporter (Bonen 2001; Galardo et al. 2007), being mostly expressed in cells with high glycolytic capacity (Bonen 2001; Bonen et al. 2006; Galardo et al. 2007). MCT4 appears to play an important role in Sertoli cells (Bonen 2001; Galardo et al. 2007; Oliveira et al. 2011, 2012; Rato et al. 2012a). Therefore, the lactate produced by Sertoli cells during their metabolic activity is exported by MCT4 and released in the intratubular fluid where it can be taken up by the developing germ cells through MCT1 and MCT2 (Oliveira et al. 2011, 2012; Mannowetz et al. 2012).

5.2   Glucose Is not the Only Fuel for the Sertoli Cell

Although glucose is a reliable substrate for energy production by Sertoli cells, other substrates can also be used for ATP synthesis. Of note is the fact that, even in the absence of glucose, Sertoli cells maintain their viability and the production of ATP (Riera et al. 2009). Moreover, processes such as endocytosis and degradation of residual bodies, as well as phagocytosis of apoptotic spermatogenic cells, occur in Sertoli cells. This enables the recycling of lipids, which can be further metabolized to produce ATP (Xiong et al. 2009). Since more than 75 % of spermatogenic cells undergo apoptosis, this is undoubtedly an important mechanism for Sertoli cell metabolism. It was also reported that spermatogenesis can be compromised by inactivation of genes involved in lipid metabolism illustrating that the metabolism of lipids is crucial for the normal occurrence of spermatogenesis (Chung et al. 2001). Besides, several studies have reported the presence of several metabolic enzymes involved in the biosynthesis of n-6 polyunsaturated fatty acids in Sertoli cells (Coniglio and Sharp 1989; Huynh et al. 1991; Oulhaj et al. 1992).

The metabolic plasticity exhibited by these cells is a well-known feature. In order to sustain an elevated metabolic activity, several alternative substrates can be used as energy source by Sertoli cells (Fig. 5.1). Several aminoacids have been described as modulators of Sertoli cell metabolism. For instance, the oxidation of glutamine and glycine is known to yield much of the energy required by Sertoli cells (Grootegoed et al. 1986). The enzyme branched-chain aminoacid aminotransferase, which is responsible for the conversion of the branched-chain aminoacids (such as valine, leucine and isoleucine) to the corresponding branched-chain acids, was identified in Sertoli cells, demonstrating that these cells can actively metabolize aminoacids (Kaiser et al. 2005). Nevertheless, the exact contribution of these substrates to Sertoli cell metabolism remains unknown and further studies are needed to disclose the contribution of these substrates to the final pool of lactate produced by these cells in physiological conditions. Notably, although the Sertoli cells have insignificant capacity for oxidizing glycine to CO_2, this aminoacid is further metabolized to exert other crucial functions. For instance, glycine can be incorporated in proteins and partially converted to lipids in Sertoli cells. Moreover, it can also be converted in serine and then incorporated in phospholipids (Kaiser et al. 2005). This illustrates that the role of aminoacids metabolism in Sertoli cells goes far beyond the nutritional support and is essential to the synthesis of protein and phospholipids.

Recent reports have described that, besides lactate, Sertoli cells also produce high amounts of acetate (Alves et al. 2012). Although the exact role for the acetate produced by Sertoli cells remains unknown, it has been suggested that it can be used to sustain the high rate of lipid metabolism by germ cells. During spermatogenesis, one of the crucial processes known to occur in the developing germ cells is the high rate of lipid synthesis and remodeling (Bajpai et al. 1998a; Oresti et al. 2010). This is a process where glucose and lipid metabolism sub-products can be used for de novo synthesis of lipids. Acetate is a crossroad metabolite that can be converted to acetyl-CoA in the cytosol or in mitochondria by acetyl-CoA synthase (Crabtree et al. 1990) and has long been known as a central metabolite in the synthesis of cholesterol and other precursors essential for lipid and phospholipid anabolism. Thus, the synthesis of lipids in germ cells can be maintained at the expenses of the high rates of acetate produced by Sertoli cells (Alves et al. 2012).

The presence of endogenous substrates in Sertoli cells and the role of glycogen in the metabolism of these cells have been overlooked. When glucose is not available, glycogen is a mobilizable fuel storage that can be readily metabolized if cells possess the necessary machinery for its metabolism. Of note, it has been shown that glycogen has a crucial role during testicular development and it also acts as a modulator of germ cell survival (Villarroel-Espíndola et al. 2013). The presence of glycogen and glycogen phosphorylase activity has been reported in Sertoli cells (Leiderman and Mancini 1969; Slaughter and Means 1983), but the role of glycogen metabolism in spermatogenesis and the mechanisms responsible for the control of these processes remain unknown. Nevertheless, the contribution of glycogen metabolism to Sertoli cells is probably underestimated, particularly under abnormal physiological conditions since the works discussed clearly illustrate that these cells possess the necessary machinery to metabolize glycogen.

As a final remark, it is crucial to highlight that most of the findings discussed in this and in the subsequent subchapter, were attained using in vitro approaches that can hamper the complexity of Sertoli cell metabolism. Culture conditions may be somewhat different between studies and the media used for culture are very rich in glucose and aminoacids, which exert a stoichiometric pressure towards some metabolic pathways in detriment of other rendering virtually impossible to make definitive assumptions of the physiological situation.

5.3   Metabolic Support of Germ Cells and Spermatozoa

Germ cells have specific nutritional requirements during spermatogenesis and their metabolic profile suffers several alterations during their development (Bajpai et al. 1998b). The reason why this happens is unclear. One possible explanation can be related with the compartmentalization that occurs in the testicle, which restricts the access of germ cells to substances and thus, they need to modulate their metabolism (Setchell 2004). In addition, the testicle has been reported as a naturally oxygen-deprived organ (Wenger and Katschinski 2005) and oxygen can control the metabolic behavior of cells. Thus, together, these facts may explain why germ cells use different metabolic pathways for the production of energy in their different stages of development (Bajpai et al. 1998b). Sperm cells also exhibit a great metabolic flexibility concerning the metabolites they can use as fuel. Indeed, they can use different metabolic pathways to assure the production of ATP. As refereed, it is unclear why this happens but it has been suggested that this property is essential for sperm survival in the female genital tracts, since the metabolic substrates available may be very distinct and specific (Ruiz-Pesini et al. 2007).

Developing germ cells strictly depend on carbohydrate metabolism, including both aerobic and anaerobic pathways (Fig. 5.2) (Bajpai et al. 1998b). Spermatogonia lie in the basal compartment of the blood-testis barrier and are supplied with nutrient from plasma components, using glucose as fuel for ATP synthesis (Boussouar and Benahmed 2004). Spermatocytes are intermediate developing germ cells that also depend on glycolysis, although the use of lactate by these developmental stage cells has also been described, especially those that lie closer to the adluminal compartment (Galardo et al. 2007).

Spermatozoa metabolism has several specific characteristics. Two of the most important are related to the substrate of preference and the compartmentalization of their metabolism. Spermatozoa rely in fructose or glucose for their energy metabolism in such a way that they present a higher glycolytic activity and lower Krebs cycle activity than all germ cells (Fig. 5.3) (Bajpai et al. 1998b). Their metabolism is also well compartmentalized, since oxidative phosphorylation occurs in the midpiece where mitochondria are located, while glycolysis occurs in the principal piece (Gomez et al. 2009). Thus, spermatozoa metabolism is a highly regulated process. Any alteration in the ability of spermatozoa to metabolize these substrates and produce ATP compromises sperm quality and thus, male fertility (Lin et al. 2009). The process of fertilization is very complex and involves a species-specific interaction between egg and sperm that leads to the formation of a viable zygote that will develop into a fetus. However, sperm must undergo a

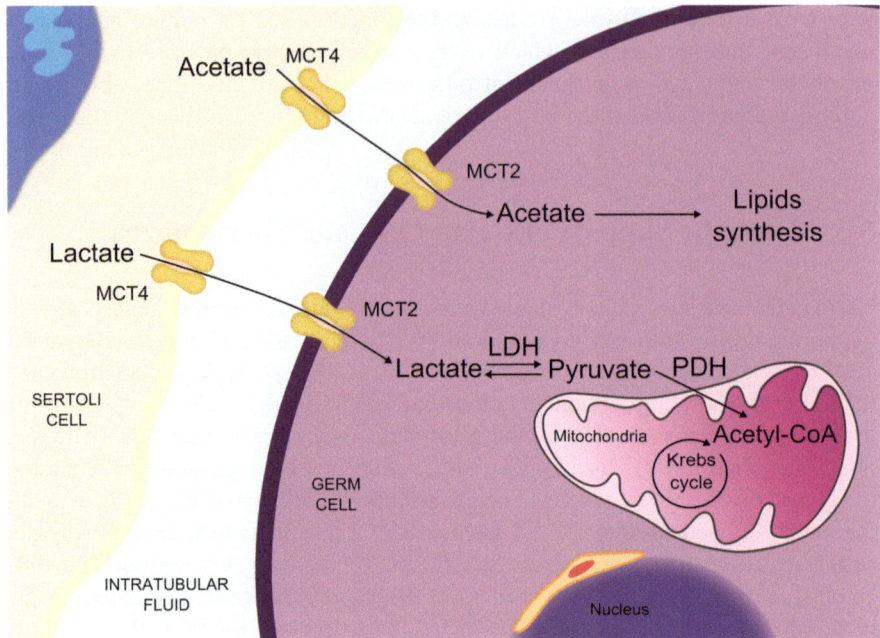

Fig. 5.2 Schematic diagram of mechanisms of metabolic cooperation between Sertoli cells and developing germ cells. Lactate and acetate are produced and exported by Sertoli cell to the intra-tubular fluid by specific monocarboxylate transporters (*MCT4*). These substrates are taken up by developing germ cells by specific monocarboxylate transporters (*MCT2*). In germ cells, acetate is used for lipid synthesis, and lactate is converted by lactate dehydrogenase (*LDH*) into pyruvate. Pyruvate is then converted by pyruvate dehydrogenase (*PDH*) into acetyl-CoA, which enters the mitochondria to be used in the Krebs cycle. Abbreviations: *MCT2* and *MCT4* monocarboxylate transporter 2 and 4; *LDH* lactate dehydrogenase; *PDH* pyruvate dehydrogenase

series of biochemical events that will determine its ability to bind and penetrate into the oocyte. This process occurs in the genital tract of the female and is known as sperm capacitation. It involves events such as alteration of sperm membrane potential, increased tyrosine phosphorylation of proteins, induction of hyperactivation and the acrosome reaction (Naz and Rajesh 2004). Most of these processes are very dependent of metabolic factors (Miki et al. 2004) illustrating that the metabolic support of spermatozoa is crucial until the formation of the zygote.

Glucose transport in testicle and spermatozoa presents a high degree of functional flexibility. The expression of the SGLT family in spermatozoa is not yet entirely characterized. For instance, it has been reported that dog spermatozoa expresses one protein of the SGLT family (Rigau et al. 2002). However the presence and distribution of SGLT in spermatozoa of other mammalians, including humans, remains unknown. The presence and distribution of GLUTs are more characterized. GLUT5 was the first GLUT member to be identified in ejaculated spermatozoa (Burant et al. 1992). It was later identified in the subequatorial region

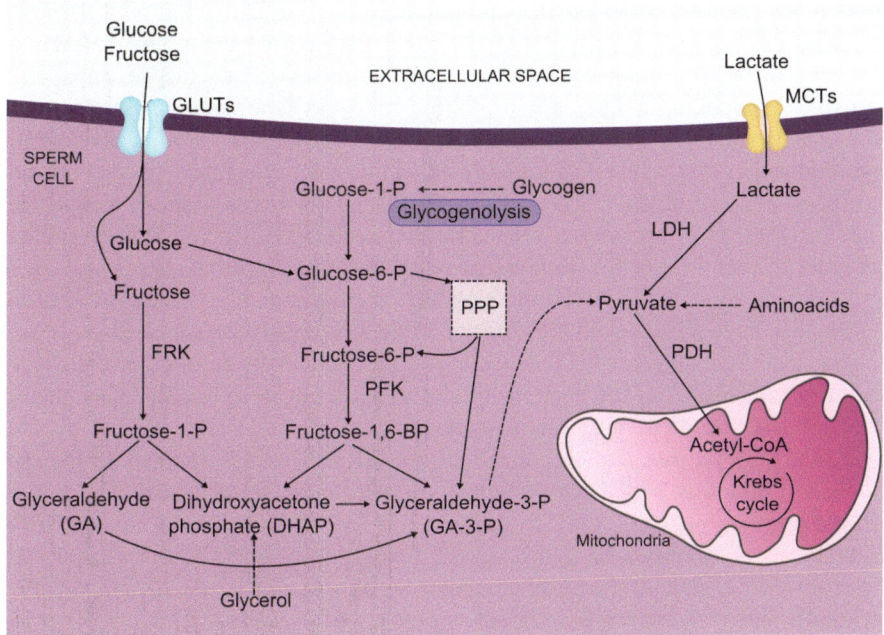

Fig. 5.3 Representative diagram of the energy metabolism pathways in sperm cells. Spermatozoa have the ability to use several substrates, namely glucose, fructose, lactate and aminoacids to obtain energy (ATP). Specific transporters, known as glucose transporters (*GLUTs*), allow the entrance of fructose and glucose through the lipidic bilayer of the cell. Glucose is converted in glucose-6-phosphate. Glycogen sources can also lead to the formation of glucose 1-phosphate, which in turn originates glucose-6-phosphate. This substrate follows through the glycolytic pathway, being converted in fructose-6-phosphate, or enters the pentose phosphate pathway (*PPP*) leading to the formation of glyceraldehyde-3-phosphate. A reaction catalyzed by phosphofructokinase (*PFK*) converts fructose-6-phosphate to fructose-1, 6-biphosphate. Fructose is converted, through fructolysis pathways, in fructose-1-phosphate by the action of fructokinase (*FRK*). Metabolites from this pathway, and from endogenous sources of glycerol, are easily converted. Both glycolysis and fructolysis converge in the formation of glyceraldehyde-3-phosphate, a key element to form pyruvate. Lactate, a substrate that enters the cytoplasm through monocarboxylate transporters (*MCTs*) can also be converted in pyruvate in a reaction catalyzed by lactate dehydrogenase (*LDH*). Pyruvate dehydrogenase (*PDH*) action allows the conversion of pyruvate in acetyl-coA, which enters the Krebs cycle within the mitochondria. In a lesser extent, spermatozoa can use free aminoacids as energy supplies for Krebs cycle. Abbreviation: *GLUTs* glucose transporters; *-P* -phosphate; *-BP* -biphosphate; *PPP* pentose phosphate pathway; *FRK* fructokinase; *PFK* phosphofructokinase; *MCTs* monocarboxylate transporters; *LDH* lactate dehydrogenase; *PDH* pyruvate dehydrogenase

of human spermatozoa, mid and principal pieces (Angulo et al. 1998). GLUT1 and GLUT2 were described in human spermatozoa acrosomal region and also in principal and end pieces (Angulo et al. 1998), while GLUT3 was identified in the midpiece of spermatozoa and in human testicle (Angulo et al. 1998; Haber et al. 1993). GLUT8 was reported to be present in human testicle and particularly in mature spermatozoa acrosomal membrane and in post acrosomal region and tail (Gomez et al. 2006). Finally, GLUT14 was identified in human testicle (Wu and Freeze 2002). The complexity in the distribution of these glucose transporters provides compelling evidence that glucose transport is a pivotal event for reproduction. In fact, glucose metabolism supports spermatozoa function and motility (Williams and Ford 2001). ATP and ADP supplies are constantly needed not only to spermatozoa motility but also for its fertilizing activity (Calamera et al. 1982). The metabolism of sperm is a very complex mechanism that follows a significant number of pathways and is dependent of an intricate signaling. It can use glycolysis or oxidative phosphorylation, together or independently (Mann 2009; Storey 2008). Under aerobic conditions, sperm cells rely on glycolysis or fructolysis to obtain their energetic needs. The glycolytic process occurs as in other cells types. In brief, glucose from the seminal fluid enters the inner compartment of the cell and follows a sequence of reactions, originating two molecules of ATP, pyruvate and NADH. Pyruvate is then converted into lactate that accumulates in the spermatozoa. Fructolysis is a very similar process that is initiated with the entry of fructose through specific GLUTs (Burant et al. 1992; Douard and Ferraris 2008). Fructose is phosphorylated by fructokinase to fructose-1-phosphate, which is then hydrolyzed by fructose-1-phosphate aldolase to form dihydroxyacetone phosphate (DHAP) and glyceraldehyde. DHAP is converted to glyceraldehyde-3-phosphate by triosephosphate isomerase or glyceraldehyde kinase. Finally, intermediates of the glycolytic pathway are obtained and pyruvate is converted to lactate (Hoskins et al. 1971; Mann 2009). Under aerobic conditions sperm cells can obtain their energy through a more effective process than glycolysis or fructolysis. The pyruvate derived from glycolysis, instead of being converted to lactate by LDH, can be oxidized and decarboxylated by pyruvate dehydrogenase (PDH) originating acetyl-CoA, which can enter the Krebs cycle. The oxidation of the substrates of this cycle is coupled with adenosine diphosphate (ADP) phosphorylation through the mitochondrial electron transport chain and the production of ATP.

Oxygen is not absolutely required to sperm metabolism and there are several factors responsible for the control of energy to sperm motility (Storey 2008). These events are very important and allow fertilization even under detrimental conditions. Intracellular lipid reserves are used by spermatozoa as endogenous substrate for ATP production (Hartree and Mann 1959; Misro and Ramya 2012). Glycerol is also used as an energy source since it can be easily converted to fructose and activate the fructolysis pathway (Mohri and Masaki 1967). The role of aminoacids cannot be disregarded, though its metabolism originates H_2O_2, which is very detrimental since spermatozoa lack peroxidase/catalase defenses and, thus, overproduction of H_2O_2 leads to loss of sperm motility (Cabrita et al. 2011).

The process of sperm entry in the oocyte also relies on a particular process of glucose metabolism known as the pentose phosphate pathway (PPP). In this process, glucose in converted to glucose-6-phosphate and then to NADPH and pentoses (Travis et al. 2001). It branches from glycolysis at early steps, when hexokinase consumes glucose-6-phosphate. Glucose-6-phosphate is then dehydrogenated by glucose-6-phosphate dehydrogenase to yield NADPH and

6-phosphogluconolactone, which is subsequently hydrolyzed by 6-phosphoglu-conolactonase to 6-phosphogluconate. The next step compromises the oxidative decarboxylation of 6-phosphogluconate by 6-phosphogluconate dehydrogenase and yields a second NADPH and ribulose-5-phosphate. In the non-oxidative phase of PPP, ribulose-5-phosphate is converted to ribose-5-phosphate by ribu-lose-5-phosphate isomerase or to xylulose-5-phosphate by ribulose-5-phosphate 3-epimerase. This non-oxidative branch is able to recruit and provide glycolytic intermediates, such as fructose-6-phosphate and glyceraldehyde-3-phosphate (Fig. 5.4).

The PPP is not only used as a pathway to obtain energy but is also used for other processes. In germ cells, PPP is used for the maintenance of the biosynthe-sis of nucleotides, production of NADPH and ribose-5-phosphate, that are neces-sary for RNA synthesis (Bajpai et al. 1998b). In addition, when Sertoli cells are incubated in the absence of hormones, PPP is not operating at its maximum rate (Grootegoed et al. 1986). However, it was noted that hormonal signals may control PPP rate and since Sertoli cells are the main hormonal targets within the testicle, it is expected that PPP may play a crucial role in the regulatory function of hor-mones in Sertoli cells metabolism and thus, in spermatogenesis. This pathway is also a valuable source of NADPH, which is required for the generation of reduced glutathione, the major scavenger of ROS. Thus, compelling evidence shows that PPP also plays a crucial role in the control of the cellular redox state. Finally, sperm survival can also be achieved by the use of glycogen and endogenous sub-strates as energy source under detrimental conditions (Travis et al. 2004).

5.4 The Central Role for Lactate in Spermatogenesis

The production of lactate by Sertoli cells is a pivotal event for the normal occur-rence of spermatogenesis. Recent works have shown that these cells can adapt their metabolic behavior to ensure an adequate production and export of lac-tate into the adluminal compartment, in which reside the developing germ cells (Oliveira et al. 2012; Riera et al. 2009). In adverse conditions resulting from a decrease in glucose levels in the extracellular milieu, the Sertoli cell increases glucose uptake to maintain lactate production by modulating GLUTs expression and by activating the AMP-activated protein kinase (AMPK), PI3K, and p38 mito-gen-activated protein kinase (MAPK)-dependent pathways (Riera et al. 2009). Deprivation of insulin, which has a crucial role in the autocrine regulation of glu-cose metabolism in cells (Aquila et al. 2005), has also been reported to alter glu-cose consumption and lactate production by Sertoli cells (Oliveira et al. 2012). Insulin deprivation not only altered the expression of genes involved in the pro-duction and export of lactate, but also induced a similar adaptation in the expres-sion of GLUTs in conditions of glucose deprivation (Oliveira et al. 2012). This illustrates that Sertoli cells adapt their transport of glucose and production of lac-tate to sustain the nutritional support of spermatogenesis.

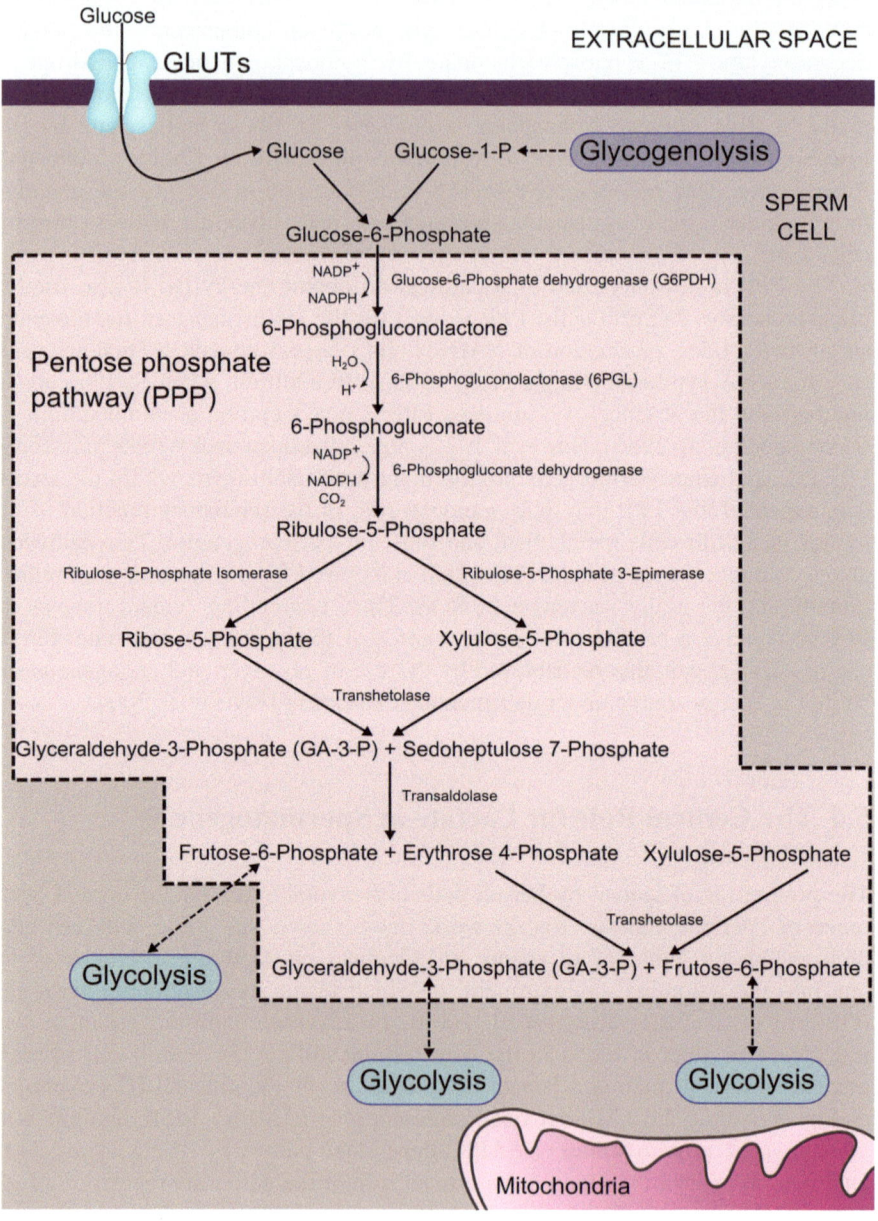

Fig. 5.4   Schematic diagram of the pentose phosphate pathway (*PPP*) in sperm cells. Glucose-6-phosphate can be obtained from two ways: from glucose taken up from extracellular space by glucose transporters (*GLUTs*), or from glycogen sources through glycogenolysis pathway, that leads to the formation of glucose 1-phosphate. In PPP, glucose-6-phosphate is dehydrogenated by glucose-6-phosphate dehydrogenase to yield NADPH and 6-phosphogluconolactone, which is subsequently hydrolyzed by 6-phosphogluconolactonase to 6-phosphogluconate. The next step compromises the oxidative decarboxylation of 6-phosphogluconate by 6-phosphogluconate dehydrogenase and yields a second NADPH and ribulose-5-phosphate. In the non-oxidative phase of PPP, ribulose-5-phosphate is converted to ribose-5-phosphate by ribulose-5-phosphate isomerase or to xylulose-5-phosphate by ribulose-5-phosphate 3-epimerase. This non-oxidative branch is able to recruit and provide glycolytic intermediates, such as fructose-6-phosphate and glyceraldehyde-3-phosphate. Abbreviations: *GLUTs* glucose transporters; *-P* -phosphate; *PPP* pentose phosphate pathway

Lactate is required, together with pyruvate, by pachytene spermatocytes and round spermatids for their energy production (Jutte et al. 1981; Mita and Hall 1982; Nakamura et al. 1982; Nakamura et al. 1984b). In in vitro studies, it was reported that pachytene spermatocytes survival and activity is regulated by the supply of lactate from Sertoli cells (Jutte et al. 1982). It has also been reported that lactate is the preferred substrate of round spermatids and that the production of energy is more efficient in the presence of high concentrations of this metabolite and low concentrations of pyruvate (Mita and Hall 1982). Yet, both lactate and pyruvate are major metabolic products with a crucial role in sustaining a high rate of protein synthesis in isolated spermatocytes and spermatids during short-term incubations (Jutte et al. 1983). In a substrate competition work, it was observed that the addition of lactate, but not glucose, enhanced respiration rates and protein and RNA synthesis stimulation by isolated pachytene spermatocytes or round spermatids (Jutte et al. 1981). Although these cells express all the enzymes of the glycolytic pathway, the energy metabolism of mature germ cells exhibits some particular characteristics since lactate is the central energy metabolite, being present in the extracellular medium and supplied by Sertoli cells (Courtens and Ploen 1999). It was also described that lactate plays a crucial role in other metabolic pathways in spermatids, as it increases protein synthesis (Nakamura et al. 1981) and modulates the rate of NADPH oxidation and the PPP (Grootegoed et al. 1986). Glucose metabolism and the metabolic sub-products related to its metabolism have other roles rather than the nutritional support of the developing germ cells. They are also important for spermatozoa motility and capacitation, since these processes are achieved through aerobic glycolysis (Miki 2007; Miki et al. 2004).

There is still a lack of direct in vivo evidence for the full understanding of the concept of Sertoli cell-germ cell metabolic cooperation. However, it is noteworthy that some in vivo models have associated alterations in spermatogenesis with the metabolism of glucose and lactate in the testicle. In fact, lactate has been reported to protect germ cells in vivo. Its importance for a normal spermatogenesis was highlighted when a study described that testicular perfusion of lactate was able to suppress the loss of spermatocytes and spermatids in adult cryptorchid rat testicle

(Courtens and Ploen 1999). Moreover, pharmacological deprivation of lactate decreases the viability of male germ cells (Trejo et al. 1995). Lactate also exerts an anti-apoptotic effect on germ cells. When human seminiferous tubules were treated with lactate, while being subjected to a protocol known to induce apoptosis, a dose-dependent suppression of germ cell death (mostly in spermatocytes and spermatids) was observed (Erkkila et al. 2002). These studies provided compelling evidence that lactate has a crucial role in the maintenance of spermatogenesis in vivo.

Throughout evolution, glycolysis is highly conserved among species. However, many glycolytic enzymes present testicle-specific isoforms and are specifically or predominantly expressed in spermatogenic cells, often during the post-meiotic phase (Gomez et al. 2009). This illustrates the relevance of glucose metabolism in testicle, particularly of the pathways responsible for the production of lactate. It is well known that lactate is produced from pyruvate through the action of LDH. The LDH enzyme family is responsible for the interconversion of pyruvate to lactate, with the concomitant oxidation/reduction of NADH to NAD^+, which is essential for the continued production of ATP by glycolysis (Kreisberg 1980). Random associations of four subunits into homo- or heterotetramers form LDH isozymes. The subunits are encoded by three loci in mammals: *Ldha*, *Ldhb* and *Ldhc*. LDHA is the predominantly expressed isoform in most tissues (Hawtrey and Goldberg 1968) while LDHC is mostly expressed in tumors (Koslowski et al. 2002) and in the testicle (Goldberg 1990). Of note, germ cells specifically express a unique isozyme of LDH (Goldberg 1985; Coonrod et al. 2006), the LDHC, which is abundant in spermatids and spermatozoa (Goldberg 1985; Goldberg et al. 2010; Li et al. 1989). Studies in mouse testicle provided evidence that the appearance of LDHC coincides and correlates with both C-subunit polypeptide (Li et al. 1989) and its mRNA (Alcivar et al. 1991), being first detectable in preleptotene spermatocytes. This unique characteristic of germ cells is related with their dependence on lactate and, as expected, the targeted disruption of the *Ldhc* gene results in male infertility due to a progressive decrease in sperm motility, failure in the capacity to develop the hyperactivated motility, a vital event for fertilization, and a decline in the levels of ATP (Odet et al. 2008). Moreover, it has been shown that LDHC escapes from transcriptional repression, resulting in significant expression levels in virtually all tumor types tested (Koslowski et al. 2002). Activation of LDHC may provide a metabolic rescue pathway in tumor cells by exploiting lactate for ATP delivery and, although germ cells are not tumor cells, it is liable that these metabolically very active cells may also use lactate for ATP delivery as they highly express this isozyme.

There are several other sperm-specific glycolytic enzymes that are encoded by genes that are only expressed during spermatogenesis. Among those we can highlight the glyceraldehyde 3-phosphate dehydrogenase-S (GAPDS) (Welch et al. 2000), phosphoglycerate kinase-2 (Boer et al. 1987; McCarrey and Thomas 1987). GAPDS is a product of a mouse gene that catalyzes the conversion of glyceraldehyde 3-phosphate to D-glycerate 1,3-biphosphate. Of note, it is only expressed during spermatogenesis and is the only glyceraldehyde 3-phosphate dehydrogenase expressed in sperm. Moreover, it is required for sperm motility (Miki et al. 2004) illustrating that the correct functioning of this enzyme is pivotal for male fertility. Indeed, studies with α-chlorohydrin and its active metabolite show that the selective inhibition of GAPDS hinders sperm glycolysis and motility in concentrations that have no effect in the isozyme in other tissues (Bone et al. 2001; Murdoch and Goldberg 2014).

Although the lactate derived from glycolysis assumes a preponderant role for spermatogenesis, the testicle exhibit a great plasticity in their metabolism and mitochondria are also identified as key players in these processes. The most abundant protein in mitochondria, the adenine nucleotide translocase (ANT), is responsible for the catalysis of the ADP/ATP exchange across the inner membrane. The ANT has multiple isoforms and one, the ANT4, has been reported as a testicle-specific isoform (Brower et al. 2007). Although the exact role for this isoform in testicular metabolism remains unknown, it has been reported that the selective disruption of *ANT4* leads to meiotic arrest at the leptotene stage during spermatogenesis (Brower et al. 2009), illustrating that the mitochondria functioning is also crucial to the normal occurrence of spermatogenesis.

5.5 Sertoli Cell Metabolism and Spermatogenesis

In sum, the data available on Sertoli cells metabolism illustrate that the catabolism of glucose and its intermediates is essential to the normal occurrence of spermatogenesis. The main metabolic events in the testicle may be summarized in: (1) the transport of glucose, which is mediated by GLUTs; (2) the reversible interconversion of pyruvate to lactate catalyzed by LDH; and (3) the transport of lactate across the plasma membrane through the action of MCTs. These events are key targets for modulating the lactate supply to developing germ cells. Moreover, germ cells must metabolize lactate in most favorable conditions and even spermatozoa maturation is dependent upon specific metabolic needs. Thus, an optimal metabolic machinery of all types of cells implicated in the process of spermatogenesis, spermiogenesis, and spermatozoa maturation, is vital for the accomplishment of an ultimate sperm quality and for fertilization. As expected, there are several factors and signals that control Sertoli cell metabolism that will be discussed in the next chapter.

Of note, most of the findings discussed were attained by using in vitro assays. Some of the events that now are assumed as correct can be altered according to the maturation state of the cells. For instance, adult mature and immature Sertoli cells have structural differences (Russell and Steinberger 1989) and they also differ in the expression of some factors such as cathepsin and transferrin (Karzai and Wright 1992). Interestingly, the hormonal response is also distinct in mature and immature Sertoli cells. Immature Sertoli cells are more responsive to FSH and less responsive to androgens (Anway et al. 2003). Moreover, there are differences in the metabolic behavior of immature and mature Sertoli cells and during culture conditions it is not clear if these cells do not revert their differentiation state. This may mask some of the findings that are now generally accepted concerning Sertoli cell metabolism. For instance, testicular cells from immature rats present a higher rate of oleate oxidation to CO_2 (Yoshida et al. 2007), illustrating that the maturational state of the cells is crucial and responsible for a distinct metabolic behavior. Finally, the number of studies using a co-culture approach between Sertoli cells and germ cells is not as high as one could expect. This kind of studies may provide more detailed information on the pathways and factors that interfere with the metabolic cooperation established between Sertoli and germ cells. Nevertheless, one should be taken into consideration that Sertoli cells isolated from the testicle of aged rats respond differently when used or not in a co-cultured system with germ cells (Syed and Hecht 2001).

The regulation of these metabolic processes is vital and could have an influence on male fertility. The modulation of metabolic pathways in testicular somatic cells, especially Sertoli cells, is likely to be determined by multiple factors including the availability of metabolic substrates and the action of hormones (Alves et al. 2012; Galardo et al. 2008; Hall and Mita 1984; Oliveira et al. 2011; Riera et al. 2001) and other endogenous or exogenous factors that are known to contribute to the progression of the spermatogenic event. The understanding of Sertoli cells energy metabolism may help to identify and support new therapeutic approaches for cases of subfertility or infertility caused by pathological conditions since spermatogenesis is in the basis of male fertility. The metabolic particularities found in the testicle and the metabolic dependence of germ cells highlight that targeting some of these features may also be a good strategy to reversibly arrest spermatogenesis and develop a male contraceptive (Murdoch and Goldberg 2014).

Chapter 6
Modulation of Sertoli Cell Metabolism

The metabolic cooperation established between testicular cells is a complex event that depends of the correct functioning of several metabolic pathways. These events are controlled through a complex network of signals that ultimately can modulate spermatogenesis. In these metabolic processes, multiple signaling cascades are triggered by hormones, proteins, metabolic products, growth factors, cytokines and many other factors. Of note, although Sertoli cell metabolism plays a crucial event to the normal occurrence of spermatogenesis, the effect of diseases and environmental pollution in the metabolic cooperation between testicular cells has not captured the attention of researchers for many years. Nevertheless, recent advances have highlighted that the metabolism of testicular cells is a target of several diseases, particularly those associated with alterations of whole body metabolic state, shedding some light over the subfertility/infertility complications that individuals suffering from those diseases exhibit. It is known that the reproductive potential of men decreases with age. For many years this outcome was attributed to cell senesce and biological aging. Nowadays it has been shown that exposure to contaminants, food intake and factors related to lifestyle may not only explain the decrease of fertility with advanced age, but also in males within reproductive age. Thus, it is crucial to discuss the factors that are responsible for the control of Sertoli cell metabolism and also how these factors are altered by diseases and exposure to toxicants that clearly induce a decrease of male fertility.

© The Author(s) 2015
P.F. Oliveira and M.G. Alves, *Sertoli Cell Metabolism and Spermatogenesis*,
SpringerBriefs in Cell Biology, DOI 10.1007/978-3-319-19791-3_6

6.1 Endogenous Factors Governing Sertoli Cell Metabolism

6.1.1 Hormones as Modulators of Sertoli Cell Metabolism

Within the seminiferous tubules, the Sertoli cell is the only testicular cell type that expresses the specific receptors for some hormones. Therefore, the Sertoli cells are considered the major hormonal targets of spermatogenesis and their functions are highly regulated by hormonal action and signaling. As expected, among the several endogenous factors that control the Sertoli cell metabolism, the hormones play a pivotal role (Fig. 6.1).

The relevance of FSH for the reproductive capacity of the males is well known. FSH controls Sertoli cell proliferation during the perinatal and/or pubertal period. Therefore, FSH determines the spermatogenic potential of the adult. FSH controls Sertoli cell metabolism via specific G-coupled receptors that are exclusively located on these cells. It has been reported that FSH stimulates spermatogenesis by enhancing the glycolytic metabolism of rat Sertoli cells, since it increases glucose uptake (Hall and Mita 1984) and both pyruvate and lactate production (Jutte et al. 1983; Riera et al. 2001). These mechanisms are thought to be highly regulated through a complex signaling network. The enhanced glucose uptake may result from the interaction between FSH and PI3K. This molecular mechanism involves the adenylyl cyclase/cAMP pathway, via activation of a G protein. FSH increases phosphorylated protein kinase B (p-PKB) levels in a PI3K-dependent mechanism (Meroni et al. 2002), which results in the translocation of GLUT1 to the plasma membrane (Samih et al. 2000). FSH also stimulates LDH activity and LDHA mRNA expression in Sertoli cells favoring the production of lactate (Riera et al. 2001).

The stimulation of lactate production and LDH activity is inhibited by wortmannin (a specific PI3K inhibitor), illustrating that the PI3K/PKB signaling pathway is a key pathway responsible for the regulation of Sertoli cell metabolism by FSH. Interestingly, FSH stimulation of glucose uptake in Sertoli cells was not altered when the cells were exposed to a protein kinase A (PKA) inhibitor illustrating that cAMP/PKA pathway is not responsible for the control of Sertoli cell metabolism by this hormone. Finally, FSH also stimulates the accumulation of aminoacids in Sertoli cells isolated from immature testicles (Wassermann et al. 1992).

TH are also known to play a crucial role in spermatogenesis and in male's fertility capacity. They regulate postnatal growth and development of the testicle, but their function throughout adult life remains a matter of debate (Sakai et al. 2004). It is well known that hyperthyroidism induces oligospermia and loss of spermatozoa in men (Clyde et al. 1976). Moreover, men with deregulation in TH are reported to have erectile dysfunction and loss of libido or even impotence (Krassas et al. 2008; Wagner et al. 2008). However, the effect of TH in Sertoli cells is not extensively studied. TH receptors were identified in rat Sertoli cells (Palmero et al. 1988) and the injection of T3 to neonatal rats completely stopped Sertoli cells proliferation (van Haaster et al. 1993). The control of Sertoli cell proliferation by TH is thought to be mediated through the action of aromatase activity and 17β-estradiol (E2) (Ando et al. 2001) and/or cyclic-dependent kinase inhibitors (Holsberger and Cooke 2005). Besides, TH are also known to alter the expression of some proteins and factors by Sertoli cells. They increase IGF1 (Palmero et al. 1990) and inhibin (van Haaster et al. 1993) but decrease ABP production (Fugassa et al. 1987) and testosterone metabolism (Palmero et al. 1995). As expected, TH are also reported to induce important changes in Sertoli cells metabolism. It was shown that cultured rat Sertoli cells, when exposed to physiological concentrations of triiodotironine, both protein synthesis and lactate production were stimulated (Palmero et al. 1995). Despite the fact that the mechanisms that mediate these processes remain undetermined, they include up-regulation of GLUT1 and modulation of MCTs and anion-transporting polypeptides (Carosa et al. 2005). Therefore, TH apparently can modulate the uptake of glucose and the export of lactate, thus altering the supply of lactate to developing germ cells. TH also stimulate the accumulation of aminoacids in immature rat testicles (Silva et al. 2001), which might be crucial to the metabolic cooperation established between testicular cells.

Sertoli cells are also targeted by sex hormones, namely androgens and estrogens. Androgens are known to be involved in the initiation and maintenance of spermatogenesis (Roberts and Zirkin 1991). In response to LH stimulation, Leydig cells produce testosterone that promotes the integrity of the blood-testis barrier and the assembly of junctional complexes (Meng et al. 2005; Wang et al. 2006). Most of the effects of androgens in Sertoli cell are mediated by androgen receptors and by 5α-reduced metabolites of testosterone, such as DHT, which present biological activities greater than testosterone (Alves et al. 2013c). As referred, androgen receptors are localized in testicular somatic cells including in Sertoli cells and Leydig cells. Germ cells from the mature testicles do not exhibit functional androgen receptors, illustrating that androgens exert their control over spermatogenesis in an indirect way by acting in the somatic cells, which directly interact with germ cells (De Gendt et al. 2004; Lyon et al. 1975; Zhou et al. 2002). Moreover, when androgen receptors are ablated from Sertoli cells, germ cell development is arrested at the spermatocyte stages (Chang et al. 2004; De Gendt et al. 2004) or early spermatid (Holdcraft and Braun 2004). Recent advances have shown that Sertoli cell metabolism is also under the control of DHT. DHT was reported to increase glucose consumption by Sertoli cells without an increase in the production of lactate. The mRNA levels of LDHA and MCT4 were found to be decreased after exposure to DHT illustrating that androgens may control Sertoli cell metabolism, particularly the production and export of lactate (Oliveira et al. 2012; Rato et al. 2012a). When exposed to flutamide, an antagonist of the androgen receptors, rat Sertoli cells also decreased the production of lactate (Goddard et al. 2003). Moreover, it was suggested that exposure to DHT shifts Sertoli cell metabolism from lactate production, as a final metabolic outcome, to Krebs cycle (Martins et al. 2013), which can compromise spermatogenesis. This is also supported by the

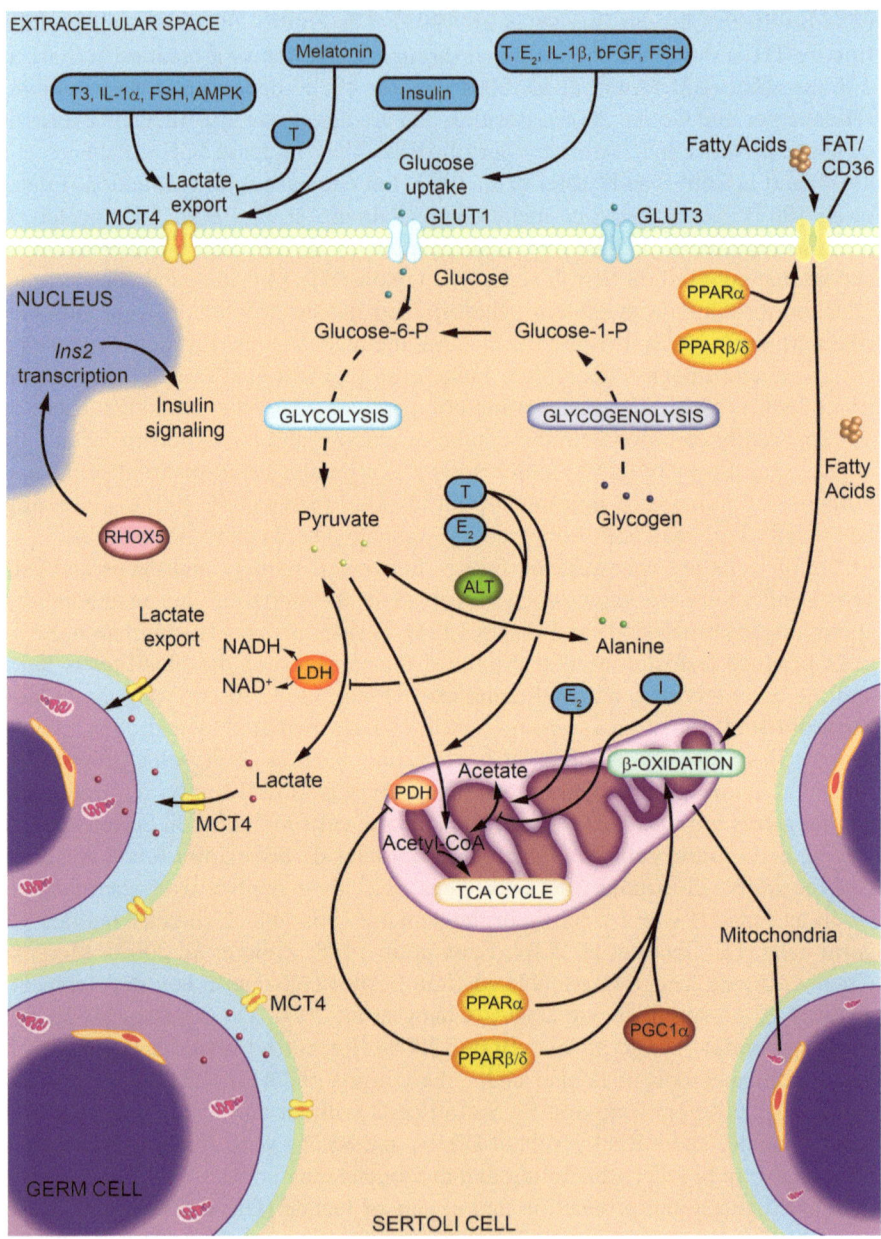

stimulatory effect of androgens in the activity of enzymes from the Krebs cycle, such as succinate and malate dehydrogenases (Gupta et al. 1991). Interestingly, the fatty acid profile of in vitro cultured Sertoli cells is also modulated by testosterone through the control of fatty acid desaturases activity (Hurtado de Catalfo and de Gomez Dumm 2005). Nevertheless, the possible role of testosterone in Sertoli cells fatty acids metabolism remains unknown and is still a matter of debate.

◀ **Fig. 6.1** Illustration of the hormonal control by endogenous factors on key metabolic pathways of Sertoli cell metabolism. The hormonal control described is mainly related to glucose and fatty acids metabolism. Sertoli cells are well placed within testicles and are continuously targeted by several hormones and other endogenous factors that modulate its metabolism. Amongst the several hormones that regulate the metabolic process, sex steroids hormones are pivotal in the control of a rate-limiting step of the glucose metabolism, the glucose uptake. Importantly, the conversion of pyruvate to lactate by lactate dehydrogenase (*LDH*) and the lactate export by monocarboxylate transporters (*MCTs*) seem to be a key point for hormonal control. Peripheral hormones, such as insulin and melatonin also play an important role in the glucose metabolism of Sertoli cells, since they coordinately regulate lactate export. Lipid metabolism is also under control and lipids are important for spermatogenic process since they act as "fuel" for Sertoli cells and are required for membrane remodeling of developing germ cells. Sertoli cells uptake fatty acids, via fatty acids transporter CD36 (*FAT/CD36*), to be oxidized in the mitochondria. Similarly to what happens with glucose, sex steroids, and peripheral hormones such as: follicle-stimulating hormone (*FSH*) and insulin also modulate lipids metabolism in these cells. Lipids oxidation is also under strict control of peroxisome proliferator-activated receptors α (*PPARα*), peroxisome proliferator-activated receptors β/δ (*PPARβ/δ*) and peroxisome proliferator-activated receptors γ coactivator 1 α (*PGC1-α*). The activation of PPARα and PPARβ/δ in Sertoli cells increase the expression of fatty acid translocase/cluster of differentiation 36 (*FAT/CD36*), favoring lipids uptake. In addition to this, PPAR activation increase the phosphorylation of acetyl-coA carboxylase resulting in decreased activity of the enzyme prompting the uptake of fatty acyl-CoA into the mitochondria for subsequent oxidation. The effects of hormonal and signaling control on Sertoli cells metabolism are signalized as: black solid lines for stimulation; blunted lines for inhibition. Abbreviations: *ALT* alanine transaminase; *AMPK* 5' adenosine monophosphate activated protein kinase; *bFGF* basic fibroblast growth factor; E_2 17-β estradiol; *FAT/CD36* fatty acid translocase/cluster of differentiation 36; *FSH* follicle-stimulating hormone; *GLUT1* glucose transporter 1; *GLUT3* glucose transporter 3; *I* Insulin, *IL-1α/β* interleukin-1α/β; *LDH* lactate dehydrogenase; *MCT4* monocarboxylate transporter; *PDH* pyruvate dehydrogenase; *PPARα* peroxisome proliferator-activated receptors α; *PPARβ/δ* peroxisome proliferator-activated receptors β/δ; *PPARγ* peroxisome proliferator-activated receptors γ; *PGC1-α* peroxisome proliferator-activated receptors γ coactivator 1 α; *RHOX5* reproductive homeobox 5; *T* testosterone; *T3* triiodothyronine; *TCA cycle* tricarboxylic cycle or Krebs cycle

The action of estrogens in male reproductive physiology involves several pathways and numerous factors due to the complexity of functions that they can modulate. The concentration of estrogens in *rete testis* fluid and the testicular interstitial fluid is significantly higher than that of systemic circulation (Rato et al. 2013; Free and Jaffe 1979). In fact, the concentration of estrogens in rat epididymis has been reported to be about 25 times higher than the levels detected in the plasma (Hess 2000; Kumari et al. 1980), illustrating that estrogens are involved in virtually all testicular functions. Although Leydig cells are known to synthesize this hormone in adult individuals, some studies suggest that Sertoli cells are the major source of estrogens in immature individuals (Carreau et al. 2009). Estrogens receptors also play a crucial role in male reproductive capacity. For instance, they are responsible for the reabsorption of the seminiferous tubules fluid before the spermatozoa enter the epididymis (Picciarelli-Lima et al. 2006). Besides, estrogens receptors are also known to mediate the direct and indirect action of estrogens in spermatogenesis (Schleicher et al. 1989). Estrogens stimulate sperm functions such as motility (Cheng and Boettcher 1979), production of lactate (Revelli et al. 1998), and metabolization of other substrates (Hicks et al. 1972). Estrogens can also induce apoptosis in sperm cells (Mishra and Shaha 2005). As expected, estrogens are also modulators of Sertoli cell metabolism (Martins et al. 2013; Oliveira et al. 2011; Rato et al. 2012a). Treatment with E2 altered the expression of glycolysis-related transporters and enzymes (Martins et al. 2013), illustrating that estrogens are important modulators of Sertoli cell metabolism.

Amongst the peripheral hormones that link the whole body metabolism and reproduction, insulin plays a key role. Specific insulin receptors have been identified in Sertoli cells (Oonk and Grootegoed 1987) and it has also been reported that insulin increases in vitro production of lactate by these cells (Oonk et al. 1985). It has been suggested that this effect is mediated by the action of this hormone through its receptors (Oonk and Grootegoed 1987). Micromolar concentrations of insulin stimulate DNA and protein synthesis as well as lactate production by Sertoli cells from immature rats (Borland et al. 1984). Furthermore, Sertoli cells cultured in conditions of insulin deprivation presented: (1) altered consumption of glucose and production of lactate; (2) altered expression of genes of enzymes and transporters involved in the production and export of lactate; and (3) an alteration of glucose uptake through modulation of GLUT1 and GLUT3 (Oliveira et al. 2012). Interestingly, a similar effect was observed in response to insulin on ovarian cells stimulated by interleukin-1 (Kol et al. 1997), which shows that insulin controls the expression of GLUTs in several types of cells. Insulin deprivation was also able to decrease caspase-dependent apoptotic signaling in culture rat Sertoli cells (Dias et al. 2013). Recently, it was described that Sertoli cells produce acetate at high rates. Although the physiological relevance for the high acetate production by Sertoli cells is still a matter of debate, as referred previously, it has been suggested that acetate may be essential for lipid synthesis and remodeling in developing germ cells (Bajpai et al. 1998a; Oresti et al. 2010). Of note, Sertoli cells cultured in conditions of insulin deprivation completely repressed the production of acetate, which may impair spermatogenesis (Alves et al. 2012). These results provide evidence that the control of Sertoli cell metabolism by insulin is also a pivotal event for spermatogenesis.

Nevertheless, the mechanisms by which insulin acts, remain largely unknown. In this regard, it was shown that reproductive homeobox 5 of the chromosome X (RHOX5) is a key mediator of insulin signaling in Sertoli cells (Maclean et al. 2013). RHOX5 is a transcriptional factor widely expressed in the testicle, particularly in Sertoli cells, which is required for insulin signaling since it directly induces *insulin 2* gene transcription in Sertoli cells (Maclean et al. 2013).

A recent work reported that insulin in combination with melatonin controls the glycolytic profile of Sertoli cells (Rocha et al. 2014). The hormone melatonin, which is primarily synthesized in the pineal gland (Hardeland et al. 1995), has been reported to promote several beneficial effects to the reproductive health of males (Reiter et al. 2013). Melatonin receptors MTNR1A and MTNR1B were identified in rat Sertoli cells (Rocha et al. 2014). Furthermore, it was shown that exposure to melatonin increased glucose consumption by Sertoli cells via modulation of GLUT1 level, but decreased LDH expression and activity, which resulted in lower lactate production. Moreover, SCs exposed to melatonin produced and accumulated less acetate than insulin-exposed cells (Rocha et al. 2014).

Thus, the hormonal control of Sertoli cells function might be more complex than initially thought and it is obvious that it may involve the coordinated action of more than one hormone. For instance, it has been reported that TH can modulate estrogen receptors content in peripuberal rat Sertoli cells (Panno et al. 1996). Moreover, it is expected that the synergistic effect of hormones may be different than a simple additive effect. Further cutting-edge studies are needed to unravel the role of hormones in the metabolic cooperation established between Sertoli cells and germ cells. It is essential to disclose the mechanisms by which hormones exert a metabolic control of spermatogenesis since hormonal deregulation is usually associated with a decrease of male fertility capacity.

6.1.2 Growth Factors, Paracrine and Autocrine Mediators also Control Sertoli Cell Metabolism

Sertoli cells are able to respond to a myriad of testicular products and sub-products that can control their metabolism via paracrine or autocrine regulation. Nevertheless, this is a field of research that still needs further investigation and only a few works were done in the last decades to unravel the complexity of the action of these factors. The FGF family consists of several members that are reported to be involved in numerous processes such as angiogenesis, embryonic development, tumor growth among others (Abuharbeid et al. 2006). In fact, the bFGF is involved in the functioning and maintenance of spermatogenesis and exerts its effects through specific receptors present in Sertoli cells (Han et al. 1993). Moreover, it has been shown that bFGF regulates multiple mechanisms in the seminiferous tubules, such as tubule angiogenesis and proliferation of immature rat Sertoli cells (Mullaney and Skinner 1992). It was also described that bFGF is a testicular germ cell product involved in the control of Sertoli cells function (Han et al. 1993), evidencing that this growth factor as a key regulator of spermatogenesis. bFGF inhibits steroidogenesis in Sertoli cells from 20 to 30-day-old rats (Brucato et al. 2002) and mediates cell-cell interactions between Sertoli cells and peritubular cells (El Ramy et al. 2005). Interestingly, bFGF increases glutathione levels in Sertoli cells (Gualtieri et al. 2009), illustrating that it may also mediate oxidative stress in these cells. Of note, the action of bFGF has been suggested to be under a complex hormonal and paracrine and/or autocrine regulation by several factors including the tumor necrosis factor-alpha (TNFα) and the interleukin-1α (Le Magueresse-Battistoni et al. 1994). As could be expected, Sertoli cell metabolism is also under the control of bFGF. It has been reported that bFGF stimulates glucose metabolism in Sertoli cells through up-regulation of GLUT1 and LDHA transcript levels, as well as, through its control over LDH activity (Riera et al. 2002). Interestingly, the mechanisms by which bFGF controls the transport of glucose and the production of lactate differ, since the first is regulated by the PI3K/PKB pathway and the second is mediated by MAPK (Riera et al. 2003). In fact, the regulation of Sertoli cell function by bFGF has been described as mediated via the p42/p44-MAPK and p38-MAPK pathways that lead to stimulation of cAMP response element-binding protein (Galardo et al. 2013).

The epidermal growth factor (EGF) is a cytokine that promotes cell proliferation and controls several functions in cells. Male EGF-null mutant mice present a normal fertility (Luetteke et al. 1999), but overexpression of EGF is reported to disturb testicular function (Wong et al. 2000), illustrating that the regulation of EGF levels may be crucial for male fertility. EGF controls several important functions of Sertoli cells, namely Sertoli cell metabolism. EGF stimulates the production of lactate and inhibits aromatization in cultured Sertoli cells from immature rats (Mallea et al. 1986). Moreover, EGF also regulates several other parameters associated with glucose metabolism, such as the mRNA expression levels of LDH (Boussouar and Benahmed 1999).

Carnitine (3-hydroxy-4-N-trimethylaminobutyrate) is present in several tissues and cells due to its functions, particularly in the transport of long-chain fatty acids, as acylcarnitines, into mitochondria. Of note, throughout the male reproductive tract, several compartments present a high concentration of L-carnitine (Brooks et al. 1974; Jeulin and Lewin 1996). Carnitine supplementation to in vitro cultured Sertoli cells altered the carbohydrate metabolism of the cells, particularly the production of lactate and pyruvate and the activity of LDH (Palmero et al. 2000). The arachidonic acid is a polyunsaturated fatty acid that has also been shown to regulate the production of lactate by Sertoli cells, through stimulation of glucose uptake and increase of LDH activity (Meroni et al. 2003).

The peritubular modifying substance (PModS) is among the several paracrine factors that can alter the physiology of testicular cells. PModS is a paracrine factor derived from the peritubular cells (Skinner and Fritz 1985, 1986) that stimulates Sertoli cell differentiation and inhibin secretion (Skinner et al. 1989; Norton and Skinner 1989; Rosselli and Skinner 1992), illustrating its relevance for male reproductive physiology. In fact, the paracrine factor PModS appears to stimulate lactate production by Sertoli cells at various stages of pubertal development (Mullaney et al. 1994).

The multifunctional cytokine interleukin-1 exists in the testicle in two functional isoforms: interleukin-1α and interleukin-1β. Interleukin-1 was reported to be produced in the seminiferous epithelium by Sertoli cells and germ cells (Khan et al. 1987). Interleukin-1α was reported to stimulate LDH expression and lactate production in cultured porcine Sertoli cells (Nehar et al. 1998). Of note, the production of interleukin-1α by Sertoli cells increases during sexual maturation illustrating a possible role for this isoform during the sexual maturation events (Gerard et al. 1991). Later, it was shown that interleukin-1β also modulates Sertoli cell metabolism by increasing glucose uptake (by regulating GLUT1 expression (Galardo et al. 2008)) and modulating the production of lactate (Riera et al. 2001).

As spermatogenesis is a continuum process that has high-energy demands, Sertoli cells must possess metabolic sensors that allow them to respond to energy fluctuations. Low energy levels switch off anabolic-pathways and turn on energy-producing pathways such as glycolysis and fatty acid oxidation. Under energy deficiency, testicular ATP levels significantly decrease and may be used for adenosine formation (Rato et al. 2014), in order to stimulate the glycolytic metabolism of Sertoli cells (Galardo et al. 2010). Cultured Sertoli cells exposed to adenosine increase the transcript levels of GLUT1, which favors glucose uptake. It was reported that under those circumstances the production of lactate was also enhanced, due to an up-regulation of LDHA and MCT4 transcript levels (Galardo et al. 2010). In addition to these effects, it was suggested that adenosine may activate the energy sensor AMPK (Galardo et al. 2010). AMPK is present within the seminiferous epithelium and in Sertoli cells. The activation of AMPK by 5-aminonoimidazole-4-carboxamide-1-β-D-ribofuranoside increases glucose transport and the production of lactate, by increasing the mRNA expression of GLUT1 and MCT4. Intriguingly, AMPK activation decreased the mRNA expression of GLUT3 and MCT1, but did not alter LDH activity (Riera et al. 2007).

Recently, it was hypothesized that peroxisome proliferator-activated receptors (PPAR)—α, β/δ and γ could control the production of lactate by Sertoli cells (Regueira et al. 2014). PPARα, PPARβ/δ, PPARγ are members of the nuclear-hormone receptor superfamily and their presence in Sertoli cells as long been suggested (Braissant et al. 1996). PPARα and PPARβ/δ are reported to function as catabolic regulators, whereas PPARγ mostly regulates anabolic lipid metabolism (Garin-Shkolnik et al. 2014). PPARβ/δ was reported to decline the activity of PDH complex, which leads to a lower conversion of acetyl-coA to pyruvate. This increases the availability of pyruvate to be converted into lactate by LDH (Regueira et al. 2014).

Lipid metabolism in Sertoli cells is also under tied control by the action of several hormones and factors (Alves et al. 2013c; Rato et al. 2012b). Lipids are important for spermatogenesis since they act as "fuel" for Sertoli cells and are required for membrane remodeling of developing germ cells. Notably, FSH modulates Sertoli cell lipid metabolism by increasing the incorporation of acetate into lipids (Guma et al. 1997). These effects were also reported to be mediated by insulin action, which is known to modulate acetate incorporation and ATP citrate lyase activity, illustrating that insulin plays an important role in fatty acid synthesis. Recently, an in vitro study demonstrated that human Sertoli cells produce large amounts of acetate and that this is under a strict hormonal control (Alves et al. 2012). This metabolite is essential in the synthesis of lipids and precursors of phospholipids. Of note, insulin signaling is vital for the production of this metabolite, since deprivation of insulin suppressed acetate production by Sertoli cells. Moreover, insulin deprivation decreased the expression of acetyl-coA hydrolase (AcoA Hyd), which may explain why acetate production is significantly reduced in these conditions (Alves et al. 2012). Moreover, it is known that insulin signaling is essential for the metabolism of Sertoli cell, since insulin absence shifts the metabolism of Sertoli cells from glycolysis to Krebs cycle, which may compromise the development of germ cells (Alves et al. 2012; Oliveira et al. 2012). Sex steroids also modulate the metabolism of lipids in Sertoli cells. The effects of DHT in the metabolism of Sertoli cells were similar to those of insulin, though not as pronounced (Alves et al. 2012). However, E2 stimulated acetate production by increasing the expression of AcoA Hyd transcript levels, illustrating that E2 action may be required for the formation of sub-products essential for the maintenance of lipid synthesis. Mature Sertoli cells efficiently convert 18 carbon polyunsaturated fatty acids (PUFAs) into 22- and 24-carbon PUFAs and express the enzymes necessary for this metabolic process in high levels, including $\Delta 5$ and $\Delta 6$ desaturases and fatty acid elongases (Saether et al. 2003). However, PUFAs synthesis may be disrupted, since cultured Sertoli cells exposed to testosterone decreased the activities of both $\Delta 5$ and $\Delta 6$ desaturases (Carosa et al. 2005). These effects are of great significance, since decreased activity of $\Delta 5$ and $\Delta 6$ desaturases limit the incorporation of long chain PUFAs into sperm membranes, which ultimately compromises sperm membrane fluidity and functionality.

As lipids are essential for germ cell structure and function, its metabolism is a pivotal event for spermatogenesis. The process of fatty acid oxidation is under strict control of PPARα, PPARβ/δ and PPARγ. These transcription factors function as sensors of fatty acids and their derivatives. Interestingly, the activation of PPARα and PPARβ/δ in Sertoli cells is known to increase the expression of fatty acid transporter CD36, favoring fatty acid uptake (Regueira et al. 2014). In addition, PPAR activation increased the phosphorylation of acetyl-coA carboxylase (ACC), resulting in decreased activity of this enzyme. This resulted in prompting the uptake of fatty acyl-CoA into the mitochondria for subsequent oxidation, where mRNA levels of carnitine palmitoyltransferase 1, long chain and medium chain dehydrogenases enzymes are also up-regulated (Regueira et al. 2014). These results evidenced that PPARα and PPARβ/δ are essential for fatty acid oxidation in Sertoli cells, but the upstream regulation of PPAR system remains elusive.

As discussed, carbohydrate (particularly glucose) and lipid metabolism in Sertoli cells are tightly controlled by a wide range of factors, from peripheral hormones to nuclear receptors. Nevertheless, the role of some hormones and factors goes far beyond the modulation of metabolism. Due to the position that Sertoli cells occupy within the testicles, it is essential to clarify the mechanisms that govern their metabolism and understand to what extent they can affect the nutritional support of the developing germ cells.

6.2 Exogenous Factors

Sertoli cells are highly susceptible to numerous toxic substances, pesticides and heavy metals. Many of these substances alter the structure of these cells or induce chromatin condensation, as well as the vacuolization of the cytoplasm (Foster et al. 1982; Li et al. 2009b). Besides, the interactions between Sertoli cells and germ cells may be disrupted by many of these substances, resulting in the detachment and premature germ cell loss (Siu et al. 2009). The role of endocrine disruptors and environmental pollutants in the physiology of Sertoli cells is a matter of concern and its effects also extend to these cells metabolism (Fig. 6.2).

Several heavy metals, such as cadmium or lead, are reported to induce morphological and biochemical alterations in Sertoli cells (Adhikari et al. 2000; Hew et al. 1993; Janecki et al. 1992). Among those, mitochondrial changes in these cells were reported after in vitro exposure to lead (Bizarro et al. 2003), illustrating that heavy metals may induce crucial alterations in Sertoli cells metabolism. Notably, after being exposed to lead acetate, Sertoli cells exhibited a dose and time-dependent increase of lactate production (Batarseh et al. 1986). Similarly, a dose-dependent increase in lactate secretion by Sertoli cells was also reported when these cells were exposed to a number of reproductive toxicants, including nitrobenzene (Allenby et al. 1990), gossypol (Monsees et al. 1998), phthalate esters (Chapin et al. 1988; Moss et al. 1988) and polychlorinated biphenyls (PBCs) (Raychoudhury et al. 2000). Increased production of lactate upon toxicant exposure illustrates stimulation of glycolysis and/or inhibition of Krebs cycle and mitochondrial respiration.

Di(2-ethylhexyl) phthalate (DEHP) is an organic compound broadly used as plasticizer. In the male reproductive tract, it causes apoptosis and loss of spermatogenic cells, leading to testicular atrophy (Park et al. 2002). The way by which DEHP acts within testicles has been associated with depletion of zinc, but testicular injuries caused by the exposure to this compound also include alterations of testosterone levels and changes in testicular enzyme activities, in particular mitochondrial enzymes (Cater et al. 1977; Foster et al. 1982; Oishi 1986). DEHP-treated rats present a reduced testicular respiratory function and an increase in testicular pyruvate content (Oishi 1990), indicating that testicular mitochondrial respiratory chain is one of the main targets of DEHP. The aberrant production of pyruvate is suggested to be a compensatory mechanism to favor the production of lactate, not only to provide a correct nutritional support of germ cells, but also to serve as an anti-apoptotic agent (Erkkila et al. 2002). DEHP is a persistent toxicant and within the body it is rapidly metabolized into mono-(2-ethylhexyl) phthalate (MEHP) that promptly targets Sertoli cells and exerts similar effects to those reported to DEHP (Thysen et al. 1990). Sertoli cells exposed to MEHP significantly increased the production of lactate in a dose-dependent manner, whereas succinate dehydrogenase activity and ATP synthesis significantly decreased, demonstrating that not only Krebs cycle, but also electron transport chain are primary targets of MEHP action in Sertoli cells (Chapin et al. 1988).

Although some of these toxics apparently share the same molecular targets, their mechanism of action and the induced effects are very different. Hence, not all reproductive toxicants enhance the production of lactate in Sertoli cells. In fact, a dose-dependent decrease in the production of lactate was observed after exposure to dichlorodiphenyltrichloroethane (DDT) (Monsees et al. 2000). DDT inhibits the effect on the cAMP response to FSH signaling, which is known to be involved in the production of lactate (Meroni et al. 2002; Riera et al. 2001), eliciting a mechanism by which this pesticide may decrease the production of lactate by Sertoli cells.

Other toxicants directly modulate testicular metabolism by regulating the biochemical processes within seminiferous epithelium cells via interaction with

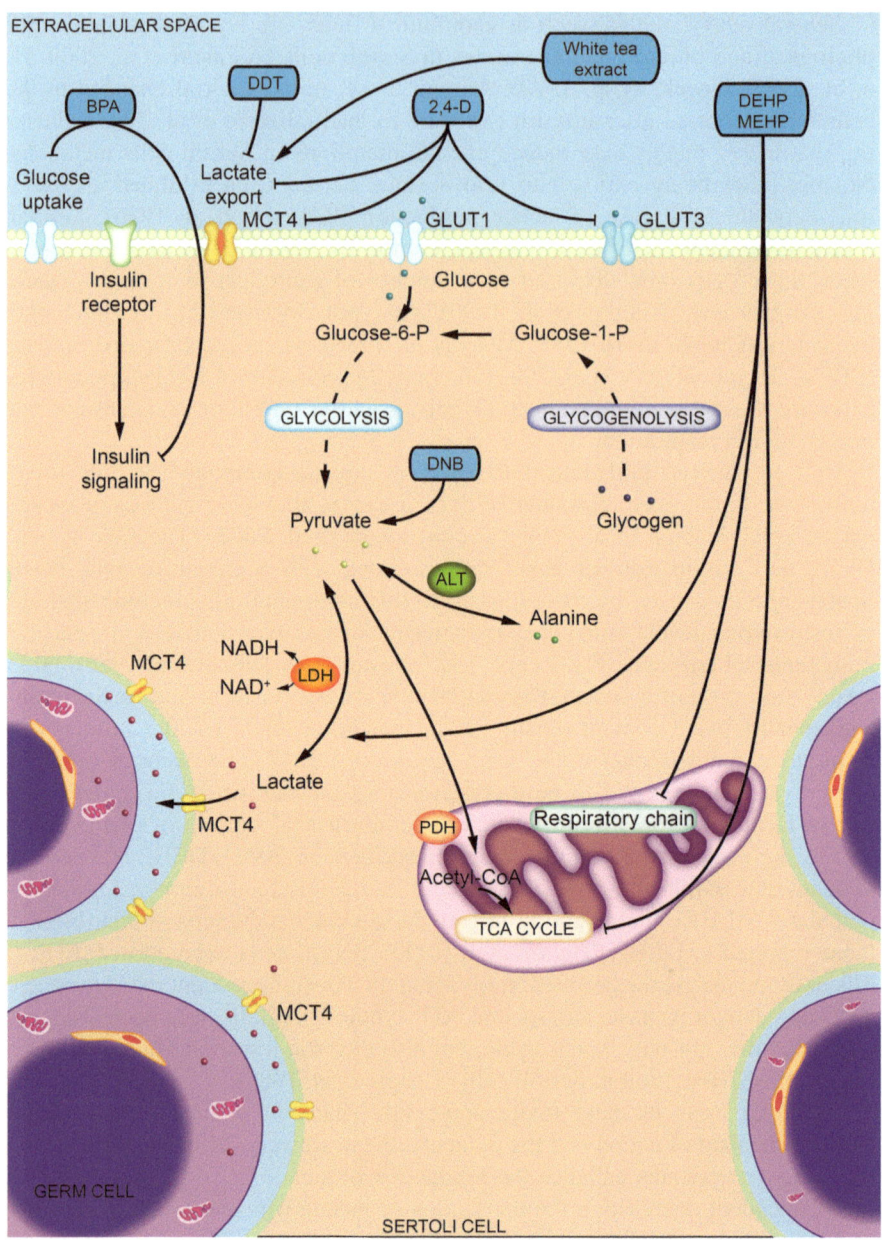

hormone receptors. In fact, many of these toxicants are classified as endocrine disruptors and thus, produce hormone-like effects. Bisphenol A (BPA), which is widely used as plasticizer, disrupts testicular insulin signaling through downregulation of insulin levels, and insulin signaling molecules, such as insulin receptor substrate 1 and 2 and PI3K (D'Cruz et al. 2012a, b). BPA compromises several

◀ **Fig. 6.2** Illustration of the control of Sertoli cell metabolism by exogenous factors. Several exogenous factors, specially the environmental toxicants which are widely used as plasticizers such as m-dinitrobenzene (*DNB*), or bisphenol A (*BPA*) alter glycolytic metabolism of Sertoli cells by increasing the production of pyruvate and lactate. Other environmental toxicants, such as Di-(2-ethylhexyl) phthalate (*DEHP*) and Mono (2-ethylhexyl) phthalate (*MEHP*), are also able to increase lactate production, and directly inhibit mitochondrial activity of Sertoli cells. Importantly, some natural products, like white tea, also enhance lactate production of these cells. Of note, some environmental toxicants may differently act, since directly interact with metabolic-related proteins by modulating their function and thus regulating the biochemical pathways, as is the case of 2,4-dichlorophenoxyacetic (2,4-D). This pesticide modulates the gene expression of transporters and enzymes associated with glycolysis. Dichlorodiphenyltrichloroethane (*DDT*) also decreases lactate export by Sertoli cells. The known effects of hormonal and signaling control on SCs metabolism is signed as: black solid lines for stimulation; blunted lines for inhibition. Abbreviations: *2,4-D* 2,4-dichlorophenoxyacetic; *ALT* alanine transaminase; *BPA* bisphenol A; *DEHP* di-(2-ethylhexyl) phthalate; *DNB* m-dinitrobenzene; *GLUT1* glucose transporter 1; *GLUT3* glucose transporter 3; *LDH* lactate dehydrogenase; *MCT4* monocarboxylate transporter; *MEHP* mono (2-ethylhexyl) phthalate, *TCA* tricarboxylic acid or Krebs cycle

insulin-dependent events, among which glucose metabolism. BPA is an estrogen-like compound able to directly interact with glucose transporters impairing the uptake of glucose into testicular cells (D'Cruz et al. 2012a). The m-dinitrobenzene (DNB) is also used in the manufacture of plastics. Exposure to DNB is also known to alter the production of lactate and pyruvate by Sertoli cells (Williams and Foster 1988). 2,4-dichlorophenoxyacetic (2,4-D) is a worldwide used pesticide classified as endocrine disruptor. It has been reported that a concentration usually detected in men working with this herbicide, is sufficient to alter the glycolytic metabolism of Sertoli cells, since the intracellular lactate levels and intracellular lactate/alanine ratio (indicator of reduced cytosolic state) are significantly decreased after exposure of these cells to 2,4-D (Alves et al. 2013b). Moreover, mRNA levels of GLUT3, PFK1, LDH and MCT4 were significantly decreased, illustrating that 2,4-D modulates genes of transporters and enzymes associated with glycolysis. When Sertoli cells were exposed to a higher dose of 2,4-D, the metabolic alterations were more pronounced as the intracellular levels of glucose, lactate and alanine were significantly decreased. These alterations were associated with a downregulation in the protein levels of GLUT3 and LDH. Since LDH is responsible for the conversion of pyruvate to lactate, the decrease in LDH levels is known to compromise the production of lactate. In addition, alanine production was also reduced and it was suggested that under these conditions Sertoli cells redirect their metabolism to the Krebs cycle, in a similar manner as reported for these cells in response to androgens. It seems that 2,4-D mimics the effects observed in cultured Sertoli cells exposed to concentrations of androgens within the physiological range (Oliveira et al. 2011).

Some natural products and pharmaceutical drugs can also alter the metabolic behavior of Sertoli cells. For instance, it has been reported that exposure to a white tea extract increases the production of lactate by Sertoli cells (Martins et al. 2014). Cisplatinum, a chemotherapeutic drug, is also known to alter the production of lactate (Malarvizhi and Mathur 1996) and the anti-diabetic drug metformin increases the production of both lactate and alanine by Sertoli cells (Alves et al. 2014).

In fact, there is compelling evidence that numerous exogenous factors can control the metabolism of Sertoli cell. Among those factors, special attention should be given to environmental toxicants, natural products and pharmaceutical drugs since exposure to these substances is widely disseminated and can promote transgenerational defects. Indeed, one of the testicular key metabolic processes (production pyruvate and/or lactate) was found to be transgenerationally altered in Sertoli cells by environmental factors (Guerrero-Bosagna et al. 2013). Thus, it is of vital importance to clarify how these external factors promote dysfunctions in the metabolic behavior of Sertoli cells.

6.3 Pathological Conditions and Sertoli Cell Metabolism—Brief Overview and the Example of Diabetes Mellitus

Several pathologies have been associated with Sertoli cell dedifferentiation. For instance, it has been reported that the seminal fluid presents reduced levels of lactate in men suffering from germ cell aplasia (also known as Sertoli cell-only syndrome) (Jain and Halder 2012), which illustrates that lactate production by Sertoli cells may be compromised probably due to the an incomplete maturation (Bar-Shira Maymon et al. 2000). Cryptorchidism, a pathological condition characterized by the failure of one or both testicles to descend into the scrotum, is also accompanied by a dedifferentiation of Sertoli cells (Nistal et al. 2013) and may also be associated with a reduced production of lactate by these cells. It has been described that cultured seminiferous tubules of cryptorchid rats secrete less lactate than scrotal testicles (Bergh et al. 1987). Moreover, the improvement of spermatogenic process after administration of exogenous lactate into cryptorchid testicles strongly supports the hypothesis that under these conditions Sertoli cells are unable to guarantee a correct nutritional provision of germ cells (Courtens and Ploen 1999).

As Sertoli cells are responsible for the transport and metabolism of glucose producing metabolic precursors essential to the development of germ cells (Robinson and Fritz 1981), a failure in these mechanisms seriously compromise spermatogenesis. In this context, the role of metabolic diseases as a factor that contributes to the decline of male fertility, namely through modulation of Sertoli cells metabolism, is under debate. Among those diseases, Diabetes Mellitus is a major public health threat that has been consistently associated with male subfertility/infertility. It is characterized by defects in insulin secretion and/or action, leading to an inability of cells to efficiently respond to its stimulation. As a result, there are several alterations in whole body metabolism that are also reflected at testicular level (Rato et al. 2013), particularly in Sertoli cells. These cells express specific receptors for insulin and, as previously discussed, it has been reported that this hormone is able to modulate the rate of lactate production by these cells

(Oonk et al. 1989). Under conditions of insulin deprivation, Sertoli cells maintain to some extent the production of lactate by regulating the expression and activity of metabolic enzymes and transporters (Oliveira et al. 2012). However, the production of acetate is severely affected (Alves et al. 2012), which may compromise the lipid synthesis and remodeling in developing germ cells. The epidemic proportion of Diabetes Mellitus is also attributed to the modern food habits. Recent reports have shown that the consumption of high-energy diets may induce prediabetes, a prodromal stage of Diabetes Mellitus in which some but not all characteristics are yet established (Tabak et al. 2012). Of note, the prediabetic state was reported to alter HCO_3^- homeodynamics in the testicle and epididymis (Bernardino et al. 2013b), as well as testicular glycolytic profile (Rato et al. 2013) and altered the testicular PGC1-α/SIRT3 axis, modulating mitochondrial bioenergetics and oxidative stress (Rato et al. 2014). Therefore, metabolic diseases, which are a major health threat, are expected to be responsible or at least significantly contribute to the declining rates of male fertility. Moreover, Sertoli cell malfunction that results from problems in their differentiation and/or state of maturation is on the basis of the reproductive dysfunction associated with those metabolic diseases. Thus, it is essential to disclose the physiology, structure and function of Sertoli cell, as well as its development and state of maturation, in men suffering pathological conditions associated with subfertility/infertility. Finally, it is crucial to unravel how all these processes can be influenced by our lifestyle and the environment to evaluate the impact of those factors in male fertility.

Chapter 7
Concluding Remarks

Spermatogenesis is a very complex process that depends on the coordinated action of the different types of testicular cells. The metabolic cooperation that is established between Sertoli cells and the developing germ cell is essential for male fertility. Indeed, the glycolytic flux of Sertoli cells is pivotal for spermatogenesis and presents some unique features. Of note, it has been proposed that these cells present a metabolic behavior very similar to cancer cells, particularly a Warburg-like metabolism. This characteristic has been overlooked over the past decades but compelling evidence show that this metabolic behavior presented by Sertoli cells is essential not only for the development of germ cells, but is also associated with the structure and maturation of the Sertoli cell. Although cell metabolism has been highlighted as a hot topic in several areas of research, it remains a neglected field of research in what concerns to male fertility and testicle biochemistry. However, new findings regarding the metabolic behavior of testicular cells in normal, pathological conditions and after exposure to toxicants and drugs revealed that spermatogenesis is highly dependent of the correct functioning of testicular cells metabolism. In fact, Sertoli cell metabolism is under a tight control of endocrine, paracrine and exogenous factors. The effects of some of these factors are already described and well reported, but it is expected that in the next decades new data will arise since several factors remain unknown. Therefore, this is a field of research that will deserve special attention from researchers all over the world in the future. Indeed, Sertoli cell metabolism is undoubtedly a control point for spermatogenesis and several pathological conditions known to induce subfertility/infertility are reported to alter factors that control the metabolism of these cells. Although more studies are needed to consolidate and expand the findings reported

© The Author(s) 2015
P.F. Oliveira and M.G. Alves, *Sertoli Cell Metabolism and Spermatogenesis*,
SpringerBriefs in Cell Biology, DOI 10.1007/978-3-319-19791-3_7

and discussed in this book, the metabolic functioning of Sertoli cell is an emerging field of research for andrology specialists interested in the understanding of the molecular mechanisms responsible for male infertility. Ultimately, those studies may highlight new therapeutic targets for the control of male fertility.

References

Abel M, Baker P, Charlton H, Monteiro A, Verhoeven G, De Gendt K, Guillou F, O'shaughnessy P (2008) Spermatogenesis and sertoli cell activity in mice lacking Sertoli cell receptors for follicle-stimulating hormone and androgen. Endocrinology 149(7):3279–3285

Abney TO (1999) The potential roles of estrogens in regulating Leydig cell development and function: a review. Steroids 64(9):610–617.

Abuharbeid S, Czubayko F, Aigner A (2006) The fibroblast growth factor-binding protein FGF-BP. Int J Biochem Cell Biol 38(9):1463–1468

Adastra KL, Frolova AI, Chi MM, Cusumano D, Bade M, Carayannopoulos MO, Moley KH (2012) Slc2a8 Deficiency in mice results in reproductive and growth impairments. Biol Reprod 87(2) (49):1–11. doi:10.1095/biolreprod.111.097675

Adhikari N, Sinha N, Saxena D (2000) Effect of lead on Sertoli–germ cell coculture of rat. Toxicol Lett 116(1):45–49

Aitken J, Fisher H (1994) Reactive oxygen species generation and human spermatozoa: the balance of benefit and risk. Bioessays 16(4):259–267

Alcivar AA, Trasler JM, Hake LE, Salehi-Ashtiani K, Goldberg E, Hecht NB (1991) DNA methylation and expression of the genes coding for lactate dehydrogenases A and C during rodent spermatogenesis. Biol Reprod 44(3):527–535

Alexander NT (1977) Immunological aspects of vasectomy. In: Boettcher B (ed) Immunological influence in human fertility. Academic Press, Sydney, pp 25–39

Allard EK, Blanchard KT, Boekelheide K (1996) Exogenous stem cell factor (SCF) compensates for altered endogenous SCF expression in 2, 5-hexanedione-induced testicular atrophy in rats. Biol Reprod 55(1):185–193

Allenby G, Sharpe RM, Foster PM (1990) Changes in Sertoli cell function in vitro induced by nitrobenzene. Fundam Appl Toxicol 14(2):364–375

Alves MG, Martins AD, Rato L, Moreira PI, Socorro S, Oliveira PF (2013a) Molecular mechanisms beyond glucose transport in diabetes-related male infertility. Biochim Biophys Acta 1832(5):626–635. doi:10.1016/j.bbadis.2013.01.011

Alves MG, Martins AD, Vaz CV, Correia S, Moreira PI, Oliveira PF, Socorro S (2014) Metformin and male reproduction: effects on sertoli cell metabolism. Br J Pharmacol 171(4):1033–1042. doi:10.1111/bph.12522

Alves MG, Neuhaus-Oliveira A, Moreira PI, Socorro S, Oliveira PF (2013) Exposure to 2,4-dichlorophenoxyacetic acid alters glucose metabolism in immature rat Sertoli cells. Reprod Toxicol 38C:81–88. doi:10.1016/j.reprotox.2013.03.005

Alves MG, Rato L, Carvalho RA, Moreira PI, Socorro S, Oliveira PF (2013) Hormonal control of Sertoli cell metabolism regulates spermatogenesis. Cell Mol Life Sci 70(5):777–793. doi:10.1007/s00018-012-1079-1

© The Author(s) 2015   75

P.F. Oliveira and M.G. Alves, *Sertoli Cell Metabolism and Spermatogenesis*, SpringerBriefs in Cell Biology, DOI 10.1007/978-3-319-19791-3

Alves MG, Socorro S, Silva J, Barros A, Sousa M, Cavaco JE, Oliveira PF (1823) In vitro cultured human Sertoli cells secrete high amounts of acetate that is stimulated by 17beta-estradiol and suppressed by insulin deprivation. Biochim Biophys Acta 8:1389–1394. doi:10.1016/j.bbamcr.2012.06.002

Ando S, Sirianni R, Forastieri P, Casaburi I, Lanzino M, Rago V, Giordano F, Giordano C, Carpino A, Pezzi V (2001) Aromatase expression in prepuberal Sertoli cells: effect of thyroid hormone. Mol Cell Endocrinol 178(1–2):11–21

Angulo C, Rauch MC, Droppelmann A, Reyes AM, Slebe JC, Delgado-Lopez F, Guaiquil VH, Vera JC, Concha II (1998) Hexose transporter expression and function in mammalian spermatozoa: cellular localization and transport of hexoses and vitamin C. J Cell Biochem 71(2):189–203

Anniballo R, Ubaldi F, Cobellis L, Sorrentino M, Rienzi L, Greco E, Tesarik J (2000) Criteria predicting the absence of spermatozoa in the Sertoli cell-only syndrome can be used to improve success rates of sperm retrieval. Hum Reprod 15(11):2269–2277

Anway MD, Folmer J, Wright WW, Zirkin BR (2003) Isolation of sertoli cells from adult rat testes: an approach to ex vivo studies of Sertoli cell function. Biol Reprod 68(3):996–1002

Anway MD, Wright WW, Zirkin BR, Korah N, Mort JS, Hermo L (2004) Expression and Localization of Cathepsin K in adult rat Sertoli Cells. Biol Reprod 70(3):562–569. doi:10.1095/biolreprod.103.018291

Aoki A, Fawcett DW (1975) Impermeability of Sertoli cell junctions to prolonged exposure to peroxidase. Andrologia 7(1):63–76

Aquila S, Gentile M, Middea E, Catalano S, Ando S (2005) Autocrine regulation of insulin secretion in human ejaculated spermatozoa. Endocrinology 146(2):552–557. doi:10.1210/en.2004-1252

Bajpai M, Gupta G, Jain SK, Setty BS (1998) Lipid metabolising enzymes in isolated rat testicular germ cells and changes associated with meiosis. Andrologia 30(6):311–315

Bajpai M, Gupta G, Setty BS (1998) Changes in carbohydrate metabolism of testicular germ cells during meiosis in the rat. Eur J Endocrinol 138(3):322–327

Bar-Shira Maymon B, Paz G, Elliott DJ, Hammel I, Kleiman SE, Yogev L, Hauser R, Botchan A, Yavetz H (2000) Maturation phenotype of Sertoli cells in testicular biopsies of azoospermic men. Hum Reprod 15(7):1537–1542

Bartlett JM, Charlton HM, Robinson IC, Nieschlag E (1990) Pubertal development and testicular function in the male growth hormone-deficient rat. J Endocrinol 126(2):193–201

Batarseh LI, Welsh MJ, Brabec MJ (1986) Effect of lead acetate on Sertoli cell lactate production and protein synthesis in vitro. Cell Biol Toxicol 2(2):283–292

Bellve AR, Cavicchia J, Millette CF, O'Brien DA, Bhatnagar Y, Dym M (1977) Spermatogenic cells of the prepubertal mouse: isolation and morphological characterization. J Cell Biol 74(1):68–85

Bellve AR, Moss SB (1983) Monoclonal antibodies as probes of reproductive mechanisms. Biol Reprod 28(1):1–26

Bergh A, Damber JE, Jacobsson H, Nilsson TK (1987) Production of lactate and tissue plasminogen activator in vitro by seminiferous tubules obtained from adult unilaterally cryptorchid rats. Arch Androl 19(2):177–182

Bernardino RL, Jesus TT, Martins AD, Sousa M, Barros A, Cavaco JE, Socorro S, Alves MG, Oliveira PF (2013) Molecular basis of bicarbonate membrane transport in the male reproductive tract. Curr Med Chem 20(32):4037–4049

Bernardino RL, Martins AD, Socorro S, Alves MG, Oliveira PF (2013) Effect of prediabetes on membrane bicarbonate transporters in testis and epididymis. J Membr Biol 246(12):877–883. doi:10.1007/s00232-013-9601-4

Besmer P, Manova K, Duttlinger R, Huang E, Packer A, Gyssler C, Bachvarova R (1993) The kit-ligand (steel factor) and its receptor c-kit/W: pleiotropic roles in gametogenesis and melanogenesis. Develop (Cambridge, England) Suppl 1993:125–137

Bizarro P, Acevedo S, Niño-Cabrera G, Mussali-Galante P, Pasos F, Avila-Costa MR, Fortoul TI (2003) Ultrastructural modifications in the mitochondrion of mouse Sertoli cells after inhalation of lead, cadmium or lead–cadmium mixture. Reprod Toxicol 17(5):561–566

Blanchard KT, Lee J, Boekelheide K (1998) Leuprolide, a gonadotropin-releasing hormone agonist, reestablishes spermatogenesis after 2, 5-hexanedione-induced irreversible testicular injury in the rat, resulting in normalized stem cell factor expression 1. Endocrinology 139(1):236–244

Boer PH, Adra CN, Lau Y, McBURNEY MW (1987) The testis-specific phosphoglycerate kinase gene pgk-2 is a recruited retroposon. Mol Cell Biol 7(9):3107–3112

Bone W, Jones AR, Morin C, Nieschlag E, Cooper TG (2001) Susceptibility of glycolytic enzyme activity and motility of spermatozoa from rat, mouse, and human to inhibition by proven and putative chlorinated antifertility compounds in vitro. J Androl 22(3):464–470

Bonen A (2001) The expression of lactate transporters (MCT1 and MCT4) in heart and muscle. Eur J Appl Physiol 86(1):6–11

Bonen A, Heynen M, Hatta H (2006) Distribution of monocarboxylate transporters MCT1-MCT8 in rat tissues and human skeletal muscle. Appl Physiol Nutr Metab 31(1):31–39. doi:10.1139/h05-002

Boockfor F, Schwarz L (1991) Effects of interleukin-6, interleukin-2, and tumor necrosis factor α on transferrin release from Sertoli Cells in culture. Endocrinology 129(1):256–262

Borland K, Mita M, Oppenheimer CL, Blinderman LA, Massague J, Hall PF, Czech MP (1984) The actions of insulin-like growth factors I and II on cultured Sertoli cells. Endocrinology 114(1):240–246

Bouffard G (1906) Injection des couleurs de benzidine aux animaux normaux. Ann Inst Pasteur (Paris) 20:539–546

Boussouar F, Benahmed M (1999) Epidermal growth factor regulates glucose metabolism through lactate dehydrogenase A messenger ribonucleic acid expression in cultured porcine Sertoli cells. Biol Reprod 61(4):1139–1145

Boussouar F, Benahmed M (2004) Lactate and energy metabolism in male germ cells. Trends Endocrinol Metab 15(7):345–350. doi:10.1016/j.tem.2004.07.003

Braissant O, Foufelle F, Scotto C, Dauca M, Wahli W (1996) Differential expression of peroxisome proliferator-activated receptors (PPARs): tissue distribution of PPAR-alpha, -beta, and -gamma in the adult rat. Endocrinology 137(1):354–366. doi:10.1210/endo.137.1.8536636

Brauchi S, Rauch MC, Alfaro IE, Cea C, Concha, II, Benos DJ, Reyes JG (2005) Kinetics, molecular basis, and differentiation of L-lactate transport in spermatogenic cells. Am J Physiol 288(3):C523–C534. 00448.2003. doi:10.1152/ajpcell.00448.2003

Brokelmann J (1963) Fine structure of germ cells and Sertoli cells during the cycle of the seminiferous epithelium in the rat. Z Zellforsch Mikrosk Anat 59:820–850

Brooks DE, Hamilton DW, Mallek AH (1974) Carnitine and glycerylphosphorylcholine in the reproductive tract of the male rat. J Reprod Fertil 36(1):141–160

Brooks JD (2007) Anatomy of the lower urinary tract and male genitalia. Campbell-Walsh Urology, 9th edn. Saunders Elsevier, Philadelphia

Brower JV, Lim CH, Jorgensen M, Oh SP, Terada N (2009) Adenine nucleotide translocase 4 deficiency leads to early meiotic arrest of murine male germ cells. Reproduction 138(3):463–470

Brower JV, Rodic N, Seki T, Jorgensen M, Fliess N, Yachnis AT, McCarrey JR, Oh SP, Terada N (2007) Evolutionarily conserved mammalian adenine nucleotide translocase 4 is essential for spermatogenesis. J Biol Chem 282(40):29658–29666

Brucato S, Bocquet J, Villers C (2002) Cell surface heparan sulfate proteoglycans: target and partners of the basic fibroblast growth factor in rat Sertoli cells. Eur J Biochem 269(2):502–511

Burant CF, Davidson NO (1994) GLUT3 glucose transporter isoform in rat testis: localization, effect of diabetes mellitus, and comparison to human testis. Am J Physiol 267(6 Pt 2): R1488–R1495

Burant CF, Takeda J, Brot-Laroche E, Bell GI, Davidson NO (1992) Fructose transporter in human spermatozoa and small intestine is GLUT5. J Biol Chem 267(21):14523–14526

Buzek SW, Sanborn BM (1988) Increase in testicular androgen receptor during sexual maturation in the rat. Biol Reprod 39(1):39–49

Buzzard JJ, Farnworth PG, De Kretser DM, O'Connor AE, Wreford NG, Morrison JR (2003) Proliferative phase Sertoli cells display a developmentally regulated response to activin in vitro. Endocrinology 144(2):474–483

Cabrita E, Ma S, Diogo P, Martínez-Páramo S, Sarasquete C, Dinis M (2011) The influence of certain aminoacids and vitamins on post-thaw fish sperm motility, viability and DNA fragmentation. Anim Reprod Sci 125(1):189–195

Calamera J, Brugo S, Vilar O (1982) Relation between motility and adenosinetriphosphate (ATP) in human spermatozoa. Andrologia 14(3):239–241

Carayannopoulos MO, Chi MM, Cui Y, Pingsterhaus JM, McKnight RA, Mueckler M, Devaskar SU, Moley KH (2000) GLUT8 is a glucose transporter responsible for insulin-stimulated glucose uptake in the blastocyst. Proc Natl Acad Sci U S A 97(13):7313–7318

Carosa E, Radico C, Giansante N, Rossi S, D'Adamo F, Di Stasi SM, Lenzi A, Jannini EA (2005) Ontogenetic profile and thyroid hormone regulation of type-1 and type-8 glucose transporters in rat Sertoli cells. Int J Androl 28(2):99–106. doi:10.1111/j.1365-2605.2005.00516.x

Carreau S, de Vienne C, Galeraud-Denis I (2008) Aromatase and estrogens in man reproduction: a review and latest advances. Adv Med Sci 53(2):139–144. doi:10.2478/v10039-008-0022-z

Carreau S, Delalande C, Galeraud-Denis I (2009) Mammalian sperm quality and aromatase expression. Microsc Res Tech 72(8):552–557. doi:10.1002/jemt.20703

Cater BR, Cook MW, Gangolli SD, Grasso P (1977) Studies on dibutyl phthalate-induced testicular atrophy in the rat: effect on zinc metabolism. Toxicol Appl Pharmacol 41(3):609–618

Cavicchia JC, Sacerdote FL, Morales A, Zhu BC (1998) Sertoli cell nuclear pore number changes in some stages of the spermatogenic cycle of the rat seminiferous epithelium. Tissue Cell 30(2):268–273

Chang C, Chen YT, Yeh SD, Xu Q, Wang RS, Guillou F, Lardy H, Yeh S (2004) Infertility with defective spermatogenesis and hypotestosteronemia in male mice lacking the androgen receptor in Sertoli cells. Proc Natl Acad Sci U S A 101(18):6876–6881. doi:10.1073/pnas.0307306101

Chapin RE, Gray TJ, Phelps JL, Dutton SL (1988) The effects of mono-(2-ethylhexyl)-phthalate on rat Sertoli cell-enriched primary cultures. Toxicol Appl Pharmacol 92(3):467–479

Chehtane M, Khaled AR (2010) Interleukin-7 mediates glucose utilization in lymphocytes through transcriptional regulation of the hexokinase II gene. Am J Physiol 298(6):C1560–C1571. doi:10.1152/ajpcell.00506.2009

Chen IL, Yates RD (1975) The fine structure and phosphatase cytochemistry of the golgi complex and associated structures in the Sertoli cells of Syrian hamsters. Cell Tissue Res 157(2):227–238

Cheng CY, Boettcher B (1979) The effect of steroids on the in vitro migration of washed human spermatozoa in modified Tyrode's solution or in fasting human blood serum. Fertil Steril 32(5):566–570

Cheng CY, Mruk DD (2002) Cell junction dynamics in the testis: Sertoli-germ cell interactions and male contraceptive development. Physiol Rev 82(4):825–874. doi:10.1152/physrev.00009.2002

Cheng CY, Mruk DD (2013) Spermatogenesis, Mammals. In: Maloy S, Hughes K (eds) Brenner's encyclopedia of genetics, 2nd edn. Academic Press, San Diego, pp 525–528. doi:10.1016/B978-0-12-374984-0.01459-5

Cheng CY, Wong EW, Yan HH, Mruk DD (2010) Regulation of spermatogenesis in the microenvironment of the seminiferous epithelium: new insights and advances. Mol Cell Endocrinol 315(1):49–56

Chiquoine AD (1964) Observations on the early events of cadmium necrosis of the testis. Anat Rec 149(1):23–35

Chung S, Wang SP, Pan L, Mitchell G, Trasler J, Hermo L (2001) Infertility and testicular defects in hormone-sensitive lipase-deficient mice. Endocrinology 142(10):4272–4281. doi:10.1210/endo.142.10.8424

Chung SS, Lee WM, Cheng CY (1999) Study on the formation of specialized inter-Sertoli cell junctions in vitro. J Cell Physiol 181(2):258–272

Chung SS, Zhu LJ, Mo MY, Silvestrini B, Lee WM, Cheng CY (1998) Evidence for cross-talk between Sertoli and germ cells using selected cathepsins as markers. J Androl 19(6):686–703

Clermont Y (1972) Kinetics of spermatogenesis in mammals: seminiferous epithelium cycle an d spermatogonial renewal. Physiol Rev 52(1):198–236

Clermont Y, Morales C, Hermo L (1987) Endocytic activities of Sertoli cells in the rat. Ann N Y Acad Sci 513:1–15

Clermont Y, Perey B (1957) Quantitative study of the cell population of the seminiferous tubules in immature rats. Am J Anat 100(2):241–267. doi:10.1002/aja.1001000205

Clulow J, Jones RC (2004) Composition of luminal fluid secreted by the seminiferous tubules and after reabsorption by the extratesticular ducts of the Japanese quail, *Coturnix coturnix japonica*. Biol Reprod 71(5):1508–1516. doi:10.1095/biolreprod.104.031401

Clyde HR, Walsh PC, English RW (1976) Elevated plasma testosterone and gonadotropin levels in infertile males with hyperthyroidism. Fertil Steril 27(6):662–666

Colvin JS, Green RP, Schmahl J, Capel B, Ornitz DM (2001) Male-to-female sex reversal in mice lacking fibroblast growth factor 9. Cell 104(6):875–889

Coniglio JG, Sharp J (1989) Biosynthesis of [14C]arachidonic acid from [14C]linoleate in primary cultures of rat Sertoli cells. Lipids 24(1):84–85

Coonrod S, Vitale A, Duan C, Bristol-Gould S, Herr J, Goldberg E (2006) Testis-specific lactate dehydrogenase (LDH-C4; Ldh3) in murine oocytes and preimplantation embryos. J Androl 27(4):502–509. doi:10.2164/jandrol.05185

Courtens JL, Ploen L (1999) Improvement of spermatogenesis in adult cryptorchid rat testis by intratesticular infusion of lactate. Biol Reprod 61(1):154–161

Crabtree B, Gordon MJ, Christie SL (1990) Measurement of the rates of acetyl-CoA hydrolysis and synthesis from acetate in rat hepatocytes and the role of these fluxes in substrate cycling. Biochem J 270(1):219–225

D'Cruz SC, Jubendradass R, Jayakanthan M, Rani SJ, Mathur PP (2012) Bisphenol A impairs insulin signaling and glucose homeostasis and decreases steroidogenesis in rat testis: an in vivo and in silico study. Food Chem Toxicol 50(3–4):1124–1133. doi:10.1016/j.fct.2011.11.041

D'Cruz SC, Jubendradass R, Mathur PP (2012) Bisphenol A induces oxidative stress and decreases levels of insulin receptor substrate 2 and glucose transporter 8 in rat testis. Reprod Sci 19(2):163–172. doi:10.1177/1933719111415547

De Gendt K, Swinnen JV, Saunders PT, Schoonjans L, Dewerchin M, Devos A, Tan K, Atanassova N, Claessens F, Lecureuil C, Heyns W, Carmeliet P, Guillou F, Sharpe RM, Verhoeven G (2004) A Sertoli cell-selective knockout of the androgen receptor causes spermatogenic arrest in meiosis. Proc Natl Acad Sci U S A 101(5):1327–1332. doi:10.1073/pnas.0308114100

De Kretser DM, Kerr J (1988) The cytology of the testis. Physiol Reprod 1:837–932

De Kretser DM, Buzzard JJ, Okuma Y, O'Connor AE, Hayashi T, Lin SY, Morrison JR, Loveland KL, Hedger MP (2004) The role of activin, follistatin and inhibin in testicular physiology. Mol Cell Endocrinol 225(1–2):57–64. doi:10.1016/j.mce.2004.07.008

De Miguel MP, De Boer-Brouwer M, Paniagua R, van den Hurk R, De Rooij DG, Van Dissel-Emiliani F (1996) Leukemia inhibitory factor and ciliary neurotropic factor promote the survival of Sertoli cells and gonocytes in coculture system. Endocrinology 137(5):1885–1893

Deslypere JP, Young M, Wilson JD, McPhaul MJ (1992) Testosterone and 5 alpha-dihydrotestosterone interact differently with the androgen receptor to enhance transcription of the MMTV-CAT reporter gene. Mol Cell Endocrinol 88(1–3):15–22

Dias TR, Rato L, Martins AD, Simões VL, Jesus TT, Alves MG, Oliveira PF (2013) Insulin deprivation decreases caspase-dependent apoptotic signaling in cultured rat Sertoli cells. ISRN Urology 2013: ID:970370

Dierich A, Sairam MR, Monaco L, Fimia GM, Gansmuller A, LeMeur M, Sassone-Corsi P (1998) Impairing follicle-stimulating hormone (FSH) signaling in vivo: targeted disruption of the FSH receptor leads to aberrant gametogenesis and hormonal imbalance. Proc Natl Acad Sci U S A 95(23):13612–13617

Dirami G, Poulter LW, Cooke BA (1991) Separation and characterization of Leydig cells and macrophages from rat testes. J Endocrinol 130(3):357–365

Doege H, Schurmann A, Bahrenberg G, Brauers A, Joost HG (2000) GLUT8, a novel member of the sugar transport facilitator family with glucose transport activity. J Biol Chem 275(21):16275–16280

Douard V, Ferraris RP (2008) Regulation of the fructose transporter GLUT5 in health and disease. Am J Physiol 295(2):E227–E237

Dym M (1994) Spermatogonial stem cells of the testis. Proc Natl Acad Sci U S A 91(24): 8346–8351

Dym M, Fawcett DW (1970) The blood-testis barrier in the rat and the physiological compartmentation of the seminiferous epithelium. Biol Reprod 3(3):308–326

Dym M, Raj HG (1977) Response of adult rat Sertoli cells and Leydig cells to depletion of luteinizing hormone and testosterone. Biol Reprod 17(5):676–696

Ebata KT, Yeh JR, Zhang X, Nagano MC (2011) Soluble growth factors stimulate spermatogonial stem cell divisions that maintain a stem cell pool and produce progenitors in vitro. Exp Cell Res 317(10):1319–1329

Eddy EM (2002) Male germ cell gene expression. Recent Prog Horm Res 57(1):103–128

El Ramy R, Verot A, Mazaud S, Odet F, Magre S, Le Magueresse-Battistoni B (2005) Fibroblast growth factor (FGF) 2 and FGF9 mediate mesenchymal-epithelial interactions of peritubular and Sertoli cells in the rat testis. J Endocrinol 187(1):135–147. doi:10.1677/joe.1.06146

Elkington JS, Fritz IB (1980) Regulation of sulfoprotein synthesis by rat Sertoli cells in culture. Endocrinology 107(4):970–976

Erkan M, Sousa M (2002) Fine structural study of the spermatogenic cycle in *Pitar rudis* and *Chamelea gallina* (Mollusca, Bivalvia, Veneridae). Tissue Cell 34(4):262–272

Erkkila K, Aito H, Aalto K, Pentikainen V, Dunkel L (2002) Lactate inhibits germ cell apoptosis in the human testis. Mol Human Reprod 8(2):109–117

Fawcett DW, Neaves WB, Flores MN (1973) Comparative observations on intertubular lymphatics and the organization of the interstitial tissue of the mammalian testis. Biol Reprod 9(5):500–532

Feig LA, Bellve AR, Erickson NH, Klagsbrun M (1980) Sertoli cells contain a mitogenic polypeptide. Proc Natl Acad Sci U S A 77(8):4774–4778

Flickinger C, Fawcett DW (1967) The junctional specializations of Sertoli cells in the seminiferous epithelium. Anat Rec 158(2):207–221. doi:10.1002/ar.1091580210

Foley GL (2001) Overview of male reproductive pathology. Toxicol Pathol 29(1):49–63

Foster P, Foster JR, Cook MW, Thomas LV, Gangolli SD (1982) Changes in ultrastructure and cytochemical localization of zinc in rat testis following the administration of Di-n-pentyl phthalate. Toxicol Appl Pharmacol 63(1):120–132

Fouquet JP (1968) Infrastructural study of the glycogen cycle in the Sertoli cells of the hamster. C R Acad Sci Hebd Seances Acad Sci D 267(5):545–548

Free MJ, Jaffe RA (1979) Collection of rete testis fluid from rats without previous efferent duct ligation. Biol Reprod 20(2):269–278

Fritz I, Tung P, Ailenberg M (1993) Proteases and antiproteases in the seminiferous tubule. In: Russell LD, Griswold MD (eds) The Sertoli cell. Cache River Press, Clearwater, pp 217–235

Fritz IB, Griswold MD, Louis BG, Dorrington JH (1976a) Similarity of responses of cultured Sertoli cells to cholera toxin and FSH. Mol Cell Endocrinol 5(3–4):289–294. 0303-7207(76)90090-3

Fritz IB, Rommerts FG, Louis BG, Dorrington JH (1976) Regulation by FSH and dibutyryl cyclic AMP of the formation of androgen-binding protein in Sertoli cell-enriched cultures. J Reprod Fertil 46(1):17–24

Fugassa E, Palmero S, Gallo G (1987) Triiodothyronine decreases the production of androgen binding protein by rat Sertoli cells. Biochem Biophys Res Commun 143(1):241–247

Fujii J, Iuchi Y, Okada F (2005) Fundamental roles of reactive oxygen species and protective mechanisms in the female reproductive system. Reprod Biol Endocrinol 3(43):1–10

Fujisawa M, Bardin CW, Morris PL (1992) A germ cell factor (s) modulates preproenkephalin gene expression in rat Sertoli cells. Mol Cell Endocrinol 84(1):79–88

Furuse M, Hirase T, Itoh M, Nagafuchi A, Yonemura S, Tsukita S (1993) Occludin: a novel integral membrane protein localizing at tight junctions. J Cell Biol 123(6):1777–1788

Galardo MN, Riera MF, Pellizzari EH, Chemes HE, Venara MC, Cigorraga SB, Meroni SB (2008) Regulation of expression of Sertoli cell glucose transporters 1 and 3 by FSH, IL1 beta, and bFGF at two different time-points in pubertal development. Cell Tissue Res 334(2):295–304. doi:10.1007/s00441-008-0656-y

Galardo MN, Riera MF, Pellizzari EH, Cigorraga SB, Meroni SB (2007) The AMP-activated protein kinase activator, 5-aminoimidazole-4-carboxamide-1-b-D-ribonucleoside, regulates lactate production in rat Sertoli cells. J Mol Endocrinol 39(4):279–288. doi:10.1677/jme-07-0054

Galardo MN, Riera MF, Pellizzari EH, Sobarzo C, Scarcelli R, Denduchis B, Lustig L, Cigorraga SB, Meroni SB (2010) Adenosine regulates Sertoli cell function by activating AMPK. Mol Cell Endocrinol 330(1–2):49–58. doi:10.1016/j.mce.2010.08.007

Galardo MN, Riera MF, Regueira M, Pellizzari EH, Cigorraga SB, Meroni SB (2013) Different signal transduction pathways elicited by basic fibroblast growth factor and interleukin 1beta regulate CREB phosphorylation in Sertoli cells. J Endocrinol Invest 36(5):331–338. doi:10.3275/8582

Garin-Shkolnik T, Rudich A, Hotamisligil GS, Rubinstein M (2014) FABP4 attenuates PPARgamma and adipogenesis and is inversely correlated with PPARgamma in adipose tissues. Diabetes 63(3):900–911. doi:10.2337/db13-0436

Gerard N, Syed V, Bardin W, Genetet N, Jegou B (1991) Sertoli cells are the site of interleukin-1 alpha synthesis in rat testis. Mol Cell Endocrinol 82(1):R13–R16

Gliki G, Ebnet K, Aurrand-Lions M, Imhof BA, Adams RH (2004) Spermatid differentiation requires the assembly of a cell polarity complex downstream of junctional adhesion molecule-C. Nature 431(7006):320–324

Goddard I, Florin A, Mauduit C, Tabone E, Contard P, Bars R, Chuzel F, Benahmed M (2003) Alteration of lactate production and transport in the adult rat testis exposed in utero to flutamide. Mol Cell Endocrinol 206(1–2):137–146

Goldberg E (1985) Reproductive implications of LDH-C4 and other testis-specific isozymes. Exp Clin Immunogenet 2(2):120–124

Goldberg E (1990) Developmental expression of lactate dehydrogenase isozymes during spermatogenesis. Prog Clin Biol Res 344:49–52

Goldberg E, Eddy EM, Duan C, Odet F (2010) LDHC: the ultimate testis-specific gene. J Androl 31(1):86–94. doi:10.2164/jandrol.109.008367

Gomez JM, Loir M, Le Gac F (1998) Growth hormone receptors in testis and liver during the spermatogenetic cycle in rainbow trout (Oncorhynchus mykiss). Biol Reprod 58(2):483–491

Gomez M, Navarro-Sabate A, Manzano A, Duran J, Obach M, Bartrons R (2009) Switches in 6-phosphofructo-2-kinase isoenzyme expression during rat sperm maturation. Biochem Biophys Res Commun 387(2):330–335. doi:10.1016/j.bbrc.2009.07.021

Gomez O, Romero A, Terrado J, Mesonero JE (2006) Differential expression of glucose transporter GLUT8 during mouse spermatogenesis. Reproduction 131(1):63–70. doi:10.1530/rep.1.00750

Griswold M, McLean D (2006) The Sertoli cell, vol 1. Knobil and Neill's Physiology of Reproduction, Elsevier

Griswold MD (1988) Protein secretions of Sertoli cells. Int Rev Cytol 110(133):133–156

Griswold MD (1995) Interactions between germ cells and Sertoli cells in the testis. Biol Reprod 52(2):211–216

Griswold MD (1998) The central role of Sertoli cells in spermatogenesis. Semin Cell Dev Biol 9(4):411–416

Griswold MD, Morales C, Sylvester SR (1988) Molecular biology of the Sertoli cell. Oxf Rev Reprod Biol 10:124–161

Grootegoed JA, Oonk RB, Jansen R, van der Molen HJ (1986) Metabolism of radiolabelled energy-yielding substrates by rat Sertoli cells. J Reprod Fertil 77(1):109–118

Grootegoed JA, Siep M, Baarends WM (2000) Molecular and cellular mechanisms in spermatogenesis. Best Practice Res Clin Endocrinol Metab 14(3):331–343

Grove B, Vogl A (1989) Sertoli cell ectoplasmic specializations: a type of actin-associated adhesion junction? J Cell Sci 93(2):309–323

Gualtieri AF, Mazzone GL, Rey RA, Schteingart HF (2009) FSH and bFGF stimulate the production of glutathione in cultured rat Sertoli cells. Int J Androl 32(3):218–225. doi:10.1111/j.1365-2605.2007.00836.x

Guerrero-Bosagna C, Savenkova M, Haque MM, Nilsson E, Skinner MK (2013) Environmentally induced epigenetic transgenerational inheritance of altered Sertoli cell transcriptome and epigenome: molecular etiology of male infertility. PLoS One 8(3):e59922. doi:10.1371/journal.pone.0059922

Guerriero G, Trocchia S, Abdel-Gawad FK, Ciarcia G (2014) Roles of reactive oxygen species in the spermatogenesis regulation. Front Endocrinol (Lausanne) 5:1–4

Guma FC, Wagner M, Martini LH, Bernard EA (1997) Effect of FSH and insulin on lipogenesis in cultures of Sertoli cells from immature rats. Braz J Med Biol Res 30(5):591–597

Guo TB, Chan KC, Hakovirta H, Xiao Y, Toppari J, Mitchell AP, Salameh WA (2003) Evidence for a role of glycogen synthase kinase-3 beta in rodent spermatogenesis. J Androl 24(3):332–342

Gupta G, Srivastava A, Setty BS (1991) Androgen-estrogen synergy in the regulation of energy metabolism in epididymis and vas deferens of rhesus monkey. Endocr Res 17(3–4):383–394

Guttenbach M, Martinez-Exposito MJ, Engel W, Schmid M (1996) Interphase chromosome arrangement in Sertoli cells of adult mice. Biol Reprod 54(5):980–986

Haber RS, Weinstein SP, O'Boyle E, Morgello S (1993) Tissue distribution of the human GLUT3 glucose transporter. Endocrinology 132(6):2538–2543

Hadley MA, Byers SW, Suárez-Quian CA, Kleinman HK, Dym M (1985) Extracellular matrix regulates Sertoli cell differentiation, testicular cord formation, and germ cell development in vitro. J Cell Biol 101(4):1511–1522

Halestrap AP (2012) The monocarboxylate transporter family—structure and functional characterization. IUBMB Life 64(1):1–9. doi:10.1002/iub.573

Hall PF, Mita M (1984) Influence of follicle-stimulating hormone on glucose transport by cultured Sertoli cells. Biol Reprod 31(5):863–869

Han IS, Sylvester SR, Kim KH, Schelling ME, Venkateswaran S, Blanckaert VD, McGuinness MP, Griswold MD (1993) Basic fibroblast growth factor is a testicular germ cell product which may regulate Sertoli cell function. Mol Endocrinol 7(7):889–897

Hardeland R, Balzer I, Poeggeler B, Fuhrberg B, Uria H, Behrmann G, Wolf R, Meyer TJ, Reiter RJ (1995) On the primary functions of melatonin in evolution: mediation of photoperiodic signals in a unicell, photooxidation, and scavenging of free radicals. J Pineal Res 18(2):104–111

Hartree E, Mann T (1959) Plasmalogen in ram semen, and its role in sperm metabolism. Biochem J 71(3):423–434

Hawtrey C, Goldberg E (1968) Differential synthesis of LDH in mouse testes. Ann N Y Acad Sci 151(1):611–615

Haywood M, Spaliviero J, Jimemez M, King NJ, Handelsman DJ, Allan CM (2003) Sertoli and germ cell development in hypogonadal (hpg) mice expressing transgenic follicle-stimulating hormone alone or in combination with testosterone. Endocrinology 144(2):509–517

Hecht N (2002) Disruption of germ cell–Sertoli cell interactions leads to spermatogenic defects. Mol Cell Endocrinol 186(2):155–157

Hecht NB (1998) Molecular mechanisms of male germ cell differentiation. Bioessays 20(7):555–561

Heckert LL, Griswold MD (2002) The expression of the follicle-stimulating hormone receptor in spermatogenesis. Recent Prog Horm Res 57:129–148

Henley JD, Young RH, Ulbright TM (2002) Malignant Sertoli cell tumors of the testis: a study of 13 examples of a neoplasm frequently misinterpreted as seminoma. Am J Surg Pathol 26(5):541–550

Hess R, Cooke PS, Bunick D, Kirby JD (1993) Adult testicular enlargement induced by neonatal hypothyroidism is accompanied by increased Sertoli and germ cell numbers. Endocrinology 132(6):2607–2613

Hess RA (2000) Oestrogen in fluid transport in efferent ducts of the male reproductive tract. Rev Reprod 5(2):84–92

Hess RA, de Franca LR (2008) Spermatogenesis and cycle of the seminiferous epithelium. In: Cheng CY (ed) Molecular mechanisms in spermatogenesis. Springer, Berlin, pp 1–15

Hess RA, França LR (2005) Structure of the Sertoli cell. In: Skinner MK, Griswold MD (eds) Sertoli Cell biology. Elsevier Academic Press, New York, pp 19–40

Hew K, Heath GL, Jiwa AH, Welsh MJ (1993) Cadmium in vivo causes disruption of tight junction-associated microfilaments in rat Sertoli cells. Biol Reprod 49(4):840–849

Heyn R, Makabe S, Motta PM (2001) Ultrastructural morphodynamics of human Sertoli cells during testicular differentiation. Ital J Anat Embryol 106(2 Suppl 2):163–171

Hicks JJ, Pedron N, Rosado A (1972) Modifications of human spermatozoa glycolysis by cyclic adenosine monophosphate (cAMP), estrogens, and follicular fluid. Fertil Steril 23(12):886–893

Hikim AS, Chakraborty J, Jhunjhunwala J (1985) Germ cell quantitation in human testicular biopsy. Urol Res 13(3):111–115

Hofmann M-C (2008) GDNF signaling pathways within the mammalian spermatogonial stem cell niche. Mol Cell Endocrinol 288(1):95–103

Holdcraft RW, Braun RE (2004) Androgen receptor function is required in Sertoli cells for the terminal differentiation of haploid spermatids. Development 131(2):459–467. doi:10.1242/dev.00957

Holsberger DR, Cooke PS (2005) Understanding the role of thyroid hormone in Sertoli cell development: a mechanistic hypothesis. Cell Tissue Res 322(1):133–140. doi:10.1007/s00441-005-1082-z

Hoskins DD, Stephens D, Casillas E (1971) Enzymic control of fructolysis in primate spermatozoa. Biochim Biophys Acta 237(2):227–238

Hu J, Chen Y-X, Wang D, Qi X, Li T-G, Hao J, Mishina Y, Garbers DL, Zhao G-Q (2004) Developmental expression and function of *Bmp4* in spermatogenesis and in maintaining epididymal integrity. Dev Biol 276(1):158–171

Hubner CA, Jentsch TJ (2002) Ion channel diseases. Hum Mol Genet 11(20):2435–2445

Hughes IA (2001) Minireview: sex differentiation. Endocrinology 142(8):3281–3287

Huleihel M, Lunenfeld E (2002) Involvement of intratesticular IL-1 system in the regulation of Sertoli cell functions. Mol Cell Endocrinol 187(1–2):125–132

Hurtado de Catalfo GE, de Gomez Dumm IN (2005) Influence of testosterone on polyunsaturated fatty acid biosynthesis in Sertoli cells in culture. Cell Biochem Funct 23(3):175–180. doi:10.1002/cbf.1135

Huynh S, Oulhaj H, Bocquet J, Nouvelot A (1991) Metabolic utilization of linoleate and alpha-linolenate in cultured Sertoli cells. Comp Biochem Physiol B 99(2):265–270

Iddon CA, Charlton H, Chiappa SA, Ponzio RO, Belis JA, Duncan JA, Reichert LJ, Abney TO, Howson JW (1977) Gonadotrophin-releasing hormone deficiency in a mutant mouse with hypogonadism. Nature 269(5626):338–340

Itoh N, Nanbu A, Tachiki H, Akagashi K, Nitta T, Mikuma N, Tsukamoto T, Kumamoto Y (1994) Restoration of Testicular transferring insulin-likegrowth factor-1 (IGF-1), and spermatogenesis by exogenously administered purified FSH and testosterone in medically hypophysectomized rats. Syst Biol Reprod Med 33(3):169–177

Jain M, Halder A (2012) Sertoli cell only syndrome: status of sertoli cell maturation and function. Ind J Endocrinol Metab 16(Suppl 2):S512–S513. doi:10.4103/2230-8210.104154

Janecki A, Jakubowiak A, Steinberger A (1992) Effect of cadmium chloride on transepithelial electrical resistance of sertoli cell monolayers in two-compartment cultures—a new model for toxicological investigations of the "blood-testis" barrier in vitro. Toxicol Appl Pharmacol 112(1):51–57

Jégou B (1992) The Sertoli cell in vivo and in vitro. Cell Biol Toxicol 8(3):49–54

Jégou B (1993) The Sertoli-germ cell communication network in mammals. Int Rev Cytol 147:25–96

Jenab S, Morris PL (1998) Testicular leukemia inhibitory factor (LIF) and LIF receptor mediate phosphorylation of signal transducers and activators of transcription (STAT)-3 and STAT-1 andinduce c-fos transcription and activator protein-1 activation in rat sertoli but not germ cells 1. Endocrinology 139(4):1883–1890

Jeulin C, Lewin LM (1996) Role of free L-carnitine and acetyl-L-carnitine in post-gonadal maturation of mammalian spermatozoa. Hum Reprod Update 2(2):87–102

Johnson L (1986) A new approach to quantification of Sertoli cells that avoids problems associated with the irregular nuclear surface. Anat Rec 214(3):231–237

Johnson L (1986) Spermatogenesis and aging in the human. J Androl 7(6):331–354

Johnson L, Petty C, Porter J, Neaves W (1984) Germ cell degeneration during postprophase of meiosis and serum concentrations of gonadotropins in young adult and older adult men. Biol Reprod 31(4):779–784

Johnson L, Thompson D (1983) Age-related and seasonal variation in the Sertoli cell population, daily sperm production and serum concentrations of follicle-stimulating hormone, luteinizing hormone and testosterone in stallions. Biol Reprod 29(3):777–789

Johnson L, Thompson DL, Jr., Varner DD (2008) Role of Sertoli cell number and function on regulation of spermatogenesis. Anim Reprod Sci 105(1–2):23–51. S0378-4320(07)00379-X [pii]. doi:10.1016/j.anireprosci.2007.11.029

Joost HG, Thorens B (2001) The extended GLUT-family of sugar/polyol transport facilitators: nomenclature, sequence characteristics, and potential function of its novel members (review). Mol Membr Biol 18(4):247–256

Juneja SC, Barr KJ, Enders GC, Kidder GM (1999) Defects in the germ line and gonads of mice lacking connexin43. Biol Reprod 60(5):1263–1270

Jutte NH, Grootegoed JA, Rommerts FF, van der Molen HJ (1981) Exogenous lactate is essential for metabolic activities in isolated rat spermatocytes and spermatids. J Reprod Fertil 62(2):399–405

Jutte NH, Jansen R, Grootegoed JA, Rommerts FF, Clausen OP, van der Molen HJ (1982) Regulation of survival of rat pachytene spermatocytes by lactate supply from Sertoli cells. J Reprod Fertil 65(2):431–438

Jutte NH, Jansen R, Grootegoed JA, Rommerts FF, van der Molen HJ (1983) FSH stimulation of the production of pyruvate and lactate by rat Sertoli cells may be involved in hormonal regulation of spermatogenesis. J Reprod Fertil 68(1):219–226

Kadam PH, Kala S, Agrawal H, Singh KP, Singh MK, Chauhan MS, Palta P, Singla SK, Manik RS (2013) Effects of glial cell line-derived neurotrophic factor, fibroblast growth factor 2 and epidermal growth factor on proliferation and the expression of some genes in buffalo (Bubalus bubalis) spermatogonial cells. Reprod Fertil Dev 25(8):1149–1157

Kaiser GR, Monteiro SC, Gelain DP, Souza LF, Perry ML, Bernard EA (2005) Metabolism of amino acids by cultured rat Sertoli cells. Metabolism 54(4):515–521. doi:10.1016/j.metabol.2004.11.005

Kanatsu-Shinohara M, Shinohara T (2013) Spermatogonial stem cell self-renewal and development. Annu Rev Cell Dev Biol 29:163–187

Karzai AW, Wright WW (1992) Regulation of the synthesis and secretion of transferrin and cyclic protein-2/cathepsin L by mature rat Sertoli cells in culture. Biol Reprod 47(5):823–831

Kerr J, Millar M, Maddocks S, Sharpe R (1993) Stage-dependent changes in spermatogenesis and sertoli cells in relation to the onset of spermatogenic failure following withdrawal of testosterone. Anat Rec 235(4):547–559

Khan SA, Soder O, Syed V, Gustafsson K, Lindh M, Ritzen EM (1987) The rat testis produces large amounts of an interleukin-1-like factor. Int J Androl 10(2):495–503

Klip A, Tsakiridis T, Marette A, Ortiz PA (1994) Regulation of expression of glucose transporters by glucose: a review of studies in vivo and in cell cultures. FASEB J 8(1):43–53

Kokk K, Verajankorva E, Wu XK, Tapfer H, Poldoja E, Pollanen P (2004) Immunohistochemical detection of glucose transporters class I subfamily in the mouse, rat and human testis. Medicina (Kaunas) 40(2):156–160

Kol S, Ben-Shlomo I, Ruutiainen K, Ando M, Davies-Hill TM, Rohan RM, Simpson IA, Adashi EY (1997) The midcycle increase in ovarian glucose uptake is associated with enhanced expression of glucose transporter 3. Possible role for interleukin-1, a putative intermediary in the ovulatory process. J Clin Invest 99(9):2274–2283. doi:10.1172/jci119403

Kopera IA, Bilinska B, Cheng CY, Mruk DD (2010) Sertoli–germ cell junctions in the testis: a review of recent data. Philos Trans R Soc B Biol Sci 365(1546):1593–1605

Koslowski M, Tureci O, Bell C, Krause P, Lehr HA, Brunner J, Seitz G, Nestle FO, Huber C, Sahin U (2002) Multiple splice variants of lactate dehydrogenase C selectively expressed in human cancer. Cancer Res 62(22):6750–6755

Kotaja N (2013) Spermatogenesis, mouse. In: Maloy S, Hughes K (eds) Brenner's encyclopedia of genetics, 2nd edn, Academic Press, San Diego, pp 529–532. doi:10.1016/B978-0-12-374984-0.01461-3

Krassas GE, Tziomalos K, Papadopoulou F, Pontikides N, Perros P (2008) Erectile dysfunction in patients with hyper- and hypothyroidism: how common and should we treat? J Clin Endocrinol Metab 93(5):1815–1819. doi:10.1210/jc.2007-2259

Kreisberg RA (1980) Lactate homeostasis and lactic acidosis. Ann Intern Med 92(2 Pt 1): 227–237

Krzanowska H, Bilinska B (2000) Number of chromocentres in the nuclei of mouse Sertoli cells in relation to the strain and age of males from puberty to senescence. J Reprod Fertil 118(2):343–350

Kumar TR, Wang Y, Lu N, Matzuk MM (1997) Follicle stimulating hormone is required for ovarian follicle maturation but not male fertility. Nat Genet 15(2):201–204

Kumari GL, Allag IS, Das RP, Datta JK (1980) Regional differences in steroidogenesis and hormone levels in the epididymis and vas deferens of adult rats. Int J Androl 3(3):267–281

Kushida T, Iijima H, Nagato Y, Kushida H (1993) Studies on thick sections of the nucleus of mouse Sertoli cells using an electron microscope operating at 300 kV. Okajimas Folia Anat Jpn 70(2–3):41–50

Le Magueresse B, Pineau C, Guillou F, Jegou B (1988) Influence of germ cells upon transferrin secretion by rat Sertoli cells in vitro. J Endocrinol 118(3):R13–R16

Le Magueresse-Battistoni B, Wolff J, Morera AM, Benahmed M (1994) Fibroblast growth factor receptor type 1 expression during rat testicular development and its regulation in cultured Sertoli cells. Endocrinology 135(6):2404–2411. doi:10.1210/endo.135.6.7988424

Lee NP, Mruk D, Lee WM, Cheng CY (2003) Is the cadherin/catenin complex a functional unit of cell-cell actin-based adherens junctions in the rat testis? Biol Reprod 68(2):489–508

Lei Z, Mishra S, Zou W, Xu B, Foltz M, Li X, Rao CV (2001) Targeted disruption of luteinizing hormone/human chorionic gonadotropin receptor gene. Mol Endocrinol 15(1):184–200

Leiderman B, Mancini RE (1969) Glycogen content in the rat testis from postnatal to adult ages. Endocrinology 85(3):607–609. doi:10.1210/endo-85-3-607

Levine E, Cupp AS, Miyashiro L, Skinner MK (2000) Role of transforming growth factor-alpha and the epidermal growth factor receptor in embryonic rat testis development. Biol Reprod 62(3):477–490

Levy S, Serre V, Hermo L, Robaire B (1999) The effects of aging on the seminiferous epithelium and the blood—testis barrier of the Brown Norway rat. J Androl 20(3):356–365

Li MW, Cheng CY, Mruk DD (2014) Sertolin mediates blood-testis barrier restructuring. Endocrinology 155(4):1520–1531

Li MW, Mruk DD, Lee WM, Cheng CY (2009) Cytokines and junction restructuring events during spermatogenesis in the testis: an emerging concept of regulation. Cytokine Growth Factor Rev 20(4):329–338

Li MW, Mruk DD, Lee WM, Cheng CY (2009) Disruption of the blood-testis barrier integrity by bisphenol A in vitro: is this a suitable model for studying blood-testis barrier dynamics? Int J Biochem Cell Biol 41(11):2302–2314. doi:10.1016/j.biocel.2009.05.016

Li SS, O'Brien DA, Hou EW, Versola J, Rockett DL, Eddy EM (1989) Differential activity and synthesis of lactate dehydrogenase isozymes A (muscle), B (heart), and C (testis) in mouse spermatogenic cells. Biol Reprod 40(1):173–180

Lin CY, Hung PH, VandeVoort CA, Miller MG (2009) 1H NMR to investigate metabolism and energy supply in rhesus macaque sperm. Reprod Toxicol 28(1):75–80. doi:10.1016/j.reprotox.2009.03.005

Lindzey J, Kumar MV, Grossman M, Young C, Tindall DJ (1994) Molecular mechanisms of androgen action. Vitam Horm 49:383–432

Luetteke NC, Qiu TH, Fenton SE, Troyer KL, Riedel RF, Chang A, Lee DC (1999) Targeted inactivation of the EGF and amphiregulin genes reveals distinct roles for EGF receptor ligands in mouse mammary gland development. Development 126(12):2739–2750

Lui W-Y, Lee WM, Cheng CY (2001) Transforming growth factor-β3 perturbs the inter-sertoli tight junction permeability barrier in vitro possibly mediated via its effects on occludin, zonula occludens-1, and claudin-11 1. Endocrinology 142(5):1865–1877

Lyon MF, Glenister PH, Lamoreux ML (1975) Normal spermatozoa from androgen-resistant germ cells of chimaeric mice and the role of androgen in spermatogenesis. Nature 258(5536):620–622

Maclean JA 2nd, Hu Z, Welborn JP, Song HW, Rao MK, Wayne CM, Wilkinson MF (2013) The RHOX homeodomain proteins regulate the expression of insulin and other metabolic regulators in the testis. J Biol Chem 288(48):34809–34825. doi:10.1074/jbc.M113.486340

Maiorino M, Ursini F (2002) Oxidative stress, spermatogenesis and fertility. Biol Chem 383(3–4):591–597

Malarvizhi D, Mathur P (1996) Effects of cisplatin on testicular functions in rats. Indian J Exp Biol 34(10):995–998

Mallea LE, Machado AJ, Navaroli F, Rommerts FF (1986) Epidermal growth factor stimulates lactate production and inhibits aromatization in cultured Sertoli cells from immature rats. Int J Androl 9(3):201–208

Mancine RE, Penhos JC, Izquierdo IA, Heinrich JJ (1960) Effects of acute hypoglycemia on rat testis. Proc Soc Exp Biol Med 104:699–702

Mann T (2009) Metabolism of semen, vol 9. Advances in enzymology and related areas of molecular biology. Wiley, Hoboken

Mannowetz N, Wandernoth P, Wennemuth G (2012) Basigin interacts with both MCT1 and MCT2 in murine spermatozoa. J Cell Physiol 227(5):2154–2162. doi:10.1002/jcp.22949

Manova K, Nocka K, Besmer P, Bachvarova R (1990) Gonadal expression of c-kit encoded at the W locus of the mouse. Development 110(4):1057–1069

Martins AD, Alves MG, Bernardino RL, Dias TR, Silva BM, Oliveira PF (2014) Effect of white tea (Camellia sinensis (L.)) extract in the glycolytic profile of Sertoli cell. Eur J Nutr 53(6):1383–1391. doi:10.1007/s00394-00013-00640-00395

Martins AD, Alves MG, Simoes VL, Dias TR, Rato L, Moreira PI, Socorro S, Cavaco JE, Oliveira PF (2013) Control of Sertoli cell metabolism by sex steroid hormones is mediated through modulation in glycolysis-related transporters and enzymes. Cell Tissue Res 354(3):861–868. doi:10.1007/s00441-013-1722-7

Marziali G, Lazzaro D, Sorrentino V (1993) Binding of germ cells to mutant SI^d sertoli cells is defective and is rescued by expression of the transmembrane form of the c-*kit* ligand. Dev Biol 157(1):182–190

Mather JP, Attie KM, Woodruff TK, Rice GC, Phillips DM (1990) Activin stimulates spermatogonial proliferation in germ-Sertoli cell cocultures from immature rat testis. Endocrinology 127(6):3206–3214

Mather JP, Moore A, Li R-H (1997) Activins, inhibins, and follistatins: further thoughts on a growing family of regulators. Exp Biol Med 215(3):209–222

McCarrey JR, Thomas K (1987) Human testis-specific PGK gene lacks introns and possesses characteristics of a processed gene. Nature 326(6112):501–505

McGinley DM, Posalaky Z, Porvaznik M, Russell L (1979) Gap junctions between Sertoli and germ cells of rat seminiferous tubules. Tissue Cell 11(4):741–754

McLachlan RI, O'Donnell L, Meachem SJ, Stanton PG, De Kretser DM, Pratis K, Robertson DM (2002) Identification of specific sites of hormonal regulation in spermatogenesis in rats, monkeys, and man. Recent Prog Horm Res 57:149–179

McLaren A (2003) Primordial germ cells in the mouse. Dev Biol 262(1):1–15

Meachem S, Von Schonfeldt V, Schlatt S (2001) Spermatogonia: stem cells with a great perspective. Reproduction 121(6):825–834

Meinhardt A, Hedger MP (2011) Immunological, paracrine and endocrine aspects of testicular immune privilege. Mol Cell Endocrinol 335(1):60–68

Meistrich M, Eng V, Loir M (1973) Temperature effects on the kinetics of spermatogenesis in the mouse. Cell Prolif 6(4):379–393

Meng J, Holdcraft RW, Shima JE, Griswold MD, Braun RE (2005) Androgens regulate the permeability of the blood-testis barrier. Proc Natl Acad Sci U S A 102(46):16696–16700. doi:10.1073/pnas.0506084102

Meng X, Lindahl M, Hyvönen ME, Parvinen M, de Rooij DG, Hess MW, Raatikainen-Ahokas A, Sainio K, Rauvala H, Lakso M (2000) Regulation of cell fate decision of undifferentiated spermatogonia by GDNF. Science 287(5457):1489–1493

Meroni SB, Riera MF, Pellizzari EH, Cigorraga SB (2002) Regulation of rat Sertoli cell function by FSH: possible role of phosphatidylinositol 3-kinase/protein kinase B pathway. J Endocrinol 174(2):195–204

Meroni SB, Riera MF, Pellizzari EH, Schteingart HF, Cigorraga SB (2003) Possible role of arachidonic acid in the regulation of lactate production in rat Sertoli cells. Int J Androl 26 (5):310–317

Middelton A (1973) Glucose metabolism in rat seminiferous tubules. University of Cambridge, England

Middleton A, Setchell BP (1972) The origin of inositol in the rete testis fluid of the ram. J Reprod Fertil 30(3):473–475

Miki K (2007) Energy metabolism and sperm function. Soc Reprod Fertility Suppl 65:309–325

Miki K, Qu W, Goulding EH, Willis WD, Bunch DO, Strader LF, Perreault SD, Eddy EM, O'Brien DA (2004) Glyceraldehyde 3-phosphate dehydrogenase-S, a sperm-specific glycolytic enzyme, is required for sperm motility and male fertility. Proc Natl Acad Sci U S A 101(47):16501–16506. doi:10.1073/pnas.0407708101

Millette CF, Bellve AR (1980) Selective partitioning of plasma membrane antigens during mouse spermatogenesis. Dev Biol 79(2):309–324

Mishra DP, Shaha C (2005) Estrogen-induced spermatogenic cell apoptosis occurs via the mitochondrial pathway: role of superoxide and nitric oxide. J Biol Chem 280(7):6181–6196. doi:10.1074/jbc.M405970200

Misro M, Ramya T (2012) Fuel/Energy Sources of Spermatozoa. In: Parekattil SJ, Agarwal A (eds) Male infertility. Springer, New York, pp 209–223

Mita M, Hall PF (1982) Metabolism of round spermatids from rats: lactate as the preferred substrate. Biol Reprod 26(3):445–455

Mital P, Hinton B, Dufour J (2011) The blood-testis and blood-epididymis barriers are more than just their tight junctions. Biol Reprod 84 (5):851-858. biolreprod.110.087452 [pii]. doi:10.1095/biolreprod.110.087452

Mohri H, Masaki J (1967) Glycerokinase and its possible role in glycerol metabolism of bull spermatozoa. J Reprod Fertil 14(2):179–194

Monsees T, Franz M, Gebhardt S, Winterstein U, Schill W, Hayatpour J (2000) Sertoli cells as a target for reproductive hazards. Andrologia 32(4–5):239

Monsees TK, Winterstein U, Schill WB, Miska W (1998) Influence of gossypol on the secretory function of cultured rat sertoli cells. Toxicon 36(5):813–816

Mor I, Cheung EC, Vousden KH (2011) Control of glycolysis through regulation of PFK1: old friends and recent additions. Cold Spring Harbor Symp Quant Biol 76:211–216. doi:10.1101/sqb.2011.76.010868

Morales C, Clermont Y, Hermo L (1985) Nature and function of endocytosis in Sertoli cells of the rat. Am J Anat 173(3):203–217

Morales C, Clermont Y, Nadler NJ (1986) Cyclic endocytic activity and kinetics of lysosomes in Sertoli cells of the rat: a morphometric analysis. Biol Reprod 34(1):207–218

Moroi S, Saitou M, Fujimoto K, Sakakibara A, Furuse M, Yoshida O, Tsukita S (1998) Occludin is concentrated at tight junctions of mouse/rat but not human/guinea pig Sertoli cells in testes. Am J Physiol 274(6 Pt 1):C1708–C1717

Moss EJ, Cook MW, Thomas LV, Gray TJ (1988) The effect of mono-(2-ethylhexyl) phthalate and other phthalate esters on lactate production by Sertoli cells in vitro. Toxicol Lett 40(1):77–84

Mruk DD, Cheng CY (2004) Sertoli-Sertoli and Sertoli-germ cell interactions and their significance in germ cell movement in the seminiferous epithelium during spermatogenesis. Endocr Rev 25 (5):747-806. 25/5/747. doi:10.1210/er.2003-0022

Mruk DD, Cheng CY (2011) Desmosomes in the testis: moving into an unchartered territory. Spermatogenesis 1(1):47–51

Mruk DD, Silvestrini B, Cheng CY (2008) Anchoring junctions as drug targets: role in contraceptive development. Pharmacol Rev 60(2):146–180

Mueller S, Rosenquist TA, Takai Y, Bronson RA, Wimmer E (2003) Loss of nectin-2 at Sertoli-spermatid junctions leads to male infertility and correlates with severe spermatozoan head and midpiece malformation, impaired binding to the zona pellucida, and oocyte penetration. Biol Reprod 69(4):1330–1340

Mullaney B, Skinner M (1993) Transforming growth factor-beta (beta 1, beta 2, and beta 3) gene expression and action during pubertal development of the seminiferous tubule: potential role at the onset of spermatogenesis. Mol Endocrinol 7(1):67–76

Mullaney BP, Rosselli M, Skinner MK (1994) Developmental regulation of Sertoli cell lactate production by hormones and the testicular paracrine factor. PModS Mol Cell Endocrinol 104(1):67–73

Mullaney BP, Skinner MK (1992) Basic fibroblast growth factor (bFGF) gene expression and protein production during pubertal development of the seminiferous tubule: follicle-stimulating hormone-induced Sertoli cell bFGF expression. Endocrinology 131(6):2928–2934. doi:10.1210/endo.131.6.1446630

Murdoch FE, Goldberg E (2014) Male contraception: another holy grail. Bioorg Med Chem Lett 24(2):419–424

Nakamura M, Fujiwara A, Yasumasu I, Okinaga S, Arai K (1982) Regulation of glucose metabolism by adenine nucleotides in round spermatids from rat testes. J Biol Chem 257(23):13945–13950

Nakamura M, Hino A, Kato J (1981) Stimulation of protein synthesis in round spermatids from rat testes by lactate. II. Role of adenosine triphosphate (ATP). J Biochem 90(4):933–940

Nakamura M, Okinaga S, Arai K (1984) Metabolism of pachytene primary spermatocytes from rat testes: pyruvate maintenance of adenosine triphosphate level. Biol Reprod 30(5):1187–1197

Nakamura M, Okinaga S, Arai K (1984) Metabolism of round spermatids: evidence that lactate is preferred substrate. Am J Physiol 247(2 Pt 1):E234–E242

Naz RK, Rajesh PB (2004) Role of tyrosine phosphorylation in sperm capacitation/acrosome reaction. Reprod Biol Endocrinol 2(1):75

Nehar D, Mauduit C, Boussouar F, Benahmed M (1998) Interleukin 1alpha stimulates lactate dehydrogenase A expression and lactate production in cultured porcine sertoli cells. Biol Reprod 59(6):1425–1432

Nilsson S, Makela S, Treuter E, Tujague M, Thomsen J, Andersson G, Enmark E, Pettersson K, Warner M, Gustafsson JA (2001) Mechanisms of estrogen action. Physiol Rev 81(4):1535–1565

Nistal M, Gonzalez-Peramato P, De Miguel MP (2013) Sertoli cell dedifferentiation in human cryptorchidism and gender reassignment shows similarities between fetal environmental and adult medical treatment estrogen and antiandrogen exposure. Reprod Toxicol 42:172–179. doi:10.1016/j.reprotox.2013.08.009

Norton JN, Skinner MK (1989) Regulation of Sertoli cell function and differentiation through the actions of a testicular paracrine factor P-Mod-S. Endocrinology 124(6):2711–2719. doi:10.1210/endo-124-6-2711

O'Brien DA, Gabel CA, Eddy EM (1993) Mouse Sertoli cells secrete mannose 6-phosphate containing glycoproteins that are endocytosed by spermatogenic cells. Biol Reprod 49(5):1055–1065

O'Donnell L, McLachlan R, Wreford N, Robertson D (1994) Testosterone promotes the conversion of round spermatids between stages VII and VIII of the rat spermatogenic cycle. Endocrinology 135(6):2608–2614

O'Donnell L, Nicholls PK, O'Bryan MK, McLachlan RI, Stanton PG (2011) Spermiation: the process of sperm release. Spermatogenesis 1(1):14–35

O'Donnell L, Robertson KM, Jones ME, Simpson ER (2001) Estrogen and spermatogenesis. Endocr Rev 22(3):289–318

O'Shaughnessy P, Verhoeven G, De Gendt K, Monteiro A, Abel M (2010) Direct action through the Sertoli cells is essential for androgen stimulation of spermatogenesis. Endocrinology 151(5):2343–2348

Oakberg EF (1956) A description of spermiogenesis in the mouse and its use in analysis of the cycle of the seminiferous epithelium and germ cell renewal. Am J Anat 99(3):391–413

Odet F, Duan C, Willis WD, Goulding EH, Kung A, Eddy EM, Goldberg E (2008) Expression of the gene for mouse lactate dehydrogenase C (Ldhc) is required for male fertility. Biol Reprod 79(1):26–34. doi:10.1095/biolreprod.108.068353

Oishi S (1986) Testicular atrophy induced by di(2-ethylhexyl)phthalate: changes in histology, cell specific enzyme activities and zinc concentrations in rat testis. Arch Toxicol 59(4):290–295

Oishi S (1990) Effects of phthalic acid esters on testicular mitochondrial functions in the rat. Arch Toxicol 64(2):143–147

Oliveira PF, Alves MG, Martins AD, Correia S, Bernardino RL, Silva J, Barros A, Sousa M, Cavaco JE, Socorro S (2014) Expression pattern of G protein-coupled receptor 30 in human seminiferous tubular cells. Gen Comp Endocrinol 201:16–20

Oliveira PF, Alves MG, Rato L, Laurentino S, Silva J, Sa R, Barros A, Sousa M, Carvalho RA, Cavaco JE (1820) Socorro S (2012) Effect of insulin deprivation on metabolism and metabolism-associated gene transcript levels of in vitro cultured human Sertoli cells. Biochim Biophys Acta 2:84–89. doi:10.1016/j.bbagen.2011.11.006

Oliveira PF, Alves MG, Rato L, Silva J, Sa R, Barros A, Sousa M, Carvalho RA, Cavaco JE, Socorro S (2011) Influence of 5alpha-dihydrotestosterone and 17beta-estradiol on human Sertoli cells metabolism. Int J Androl 34(6 Pt 2):e612–e620. doi:10.1111/j.1365-2605.2011.01205.x

Oliveira PF, Martins AD, Moreira AC, Cheng CY, Alves MG (2014) The Warburg effect revisited—lesson from the Sertoli cell. Med Res Rev:DOI. doi:10.1002/med.21325

Oliveira PF, Sousa M, Barros A, Moura T, da Costa AR (2009) Intracellular pH regulation in human Sertoli cells: role of membrane transporters. Reproduction 137(2):353–359

Oliveira PF, Sousa M, Barros A, Moura T, Rebelo da Costa A (2009) Membrane transporters and cytoplasmatic pH regulation on bovine Sertoli cells. J Membr Biol 227(1):49–55

Olson GE, Winfrey VP, NagDas SK, Hill KE, Burk RF (2005) Selenoprotein P is required for mouse sperm development. Biol Reprod 73(1):201–211

Oonk RB, Grootegoed JA (1987) Identification of insulin receptors on rat Sertoli cells. Mol Cell Endocrinol 49(1):51–62

Oonk RB, Grootegoed JA, van der Molen HJ (1985) Comparison of the effects of insulin and follitropin on glucose metabolism by Sertoli cells from immature rats. Mol Cell Endocrinol 42(1):39–48

Oonk RB, Jansen R, Grootegoed JA (1989) Differential effects of follicle-stimulating hormone, insulin, and insulin-like growth factor I on hexose uptake and lactate production by rat Sertoli cells. J Cell Physiol 139(1):210–218. doi:10.1002/jcp.1041390128

Oresti GM, Reyes JG, Luquez JM, Osses N, Furland NE, Aveldano MI (2010) Differentiation-related changes in lipid classes with long-chain and very long-chain polyenoic fatty acids in rat spermatogenic cells. J Lipid Res 51(10):2909–2921. doi:10.1194/jlr.M006429

Orth JM (1984) The role of follicle-stimulating hormone in controlling sertoli cell proliferation in testes of fetal rats. Endocrinology 115(4):1248–1255

Orth JM, Gunsalus GL, Lamperti AA (1988) Evidence from Sertoli cell-depleted rats indicates that spermatid number in adults depends on numbers of Sertoli cells produced during perinatal development. Endocrinology 122(3):787–794

Oulhaj H, Huynh S, Nouvelot A (1992) The biosynthesis of polyunsaturated fatty acids by rat sertoli cells. Comp Biochem Physiol B 102(4):897–904

Ozaki-Kuroda K, Nakanishi H, Ohta H, Tanaka H, Kurihara H, Mueller S, Irie K, Ikeda W, Sakai T, Wimmer E (2002) Nectin couples cell-cell adhesion and the actin scaffold at heterotypic testicular junctions. Curr Biol 12(13):1145–1150

Pakarainen T, Zhang F-P, Mäkelä S, Poutanen M, Huhtaniemi I (2005) Testosterone replacement therapy induces spermatogenesis and partially restores fertility in luteinizing hormone receptor knockout mice. Endocrinology 146(2):596–606

Palmero S, Bottazzi C, Costa M, Leone M, Fugassa E (2000) Metabolic effects of L-carnitine on prepubertal rat Sertoli cells. Horm Metab Res 32(3):87–90. doi:10.1055/s-2007-978596

Palmero S, Maggiani S, Fugassa E (1988) Nuclear triiodothyronine receptors in rat Sertoli cells. Mol Cell Endocrinol 58(2–3):253–256

Palmero S, Prati M, Barreca A, Minuto F, Giordano G, Fugassa E (1990) Thyroid hormone stimulates the production of insulin-like growth factor I (IGF-I) by immature rat Sertoli cells. Mol Cell Endocrinol 68(1):61–65

Palmero S, Prati M, Bolla F, Fugassa E (1995) Tri-iodothyronine directly affects rat Sertoli cell proliferation and differentiation. J Endocrinol 145(2):355–362

Panno ML, Sisci D, Salerno M, Lanzino M, Mauro L, Morrone EG, Pezzi V, Palmero S, Fugassa E, Ando S (1996) Effect of triiodothyronine administration on estrogen receptor contents in peripuberal Sertoli cells. Eur J Endocrinol 134(5):633–638

Papaioannou MD, Pitetti J-L, Ro S, Park C, Aubry F, Schaad O, Vejnar CE, Kühne F, Descombes P, Zdobnov EM (2009) Sertoli cell Dicer is essential for spermatogenesis in mice. Dev Biol 326(1):250–259

Park JD, Habeebu SS, Klaassen CD (2002) Testicular toxicity of di-(2-ethylhexyl)phthalate in young Sprague-Dawley rats. Toxicology 171(2–3):105–115

Parreira GG, Melo RC, Russell LD (2002) Relationship of sertoli-sertoli tight junctions to ectoplasmic specialization in conventional and en face views. Biol Reprod 67(4):1232–1241

Pellegrini M, Grimaldi P, Rossi P, Geremia R, Dolci S (2003) Developmental expression of BMP4/ALK3/SMAD5 signaling pathway in the mouse testis: a potential role of BMP4 in spermatogonia differentiation. J Cell Sci 116(16):3363–3372

Pellerin L (2003) Lactate as a pivotal element in neuron-glia metabolic cooperation. Neurochem Int 43(4–5):331–338

Pelletier G, El-Alfy M (2000) Immunocytochemical localization of estrogen receptors alpha and beta in the human reproductive organs. J Clin Endocrinol Metab 85(12):4835–4840

Perryman KJ, Stanton PG, Loveland KL, McLachlan R, Robertson D (1996) Hormonal dependency of neural cadherin in the binding of round spermatids to Sertoli cells in vitro. Endocrinology 137(9):3877–3883

Picciarelli-Lima P, Oliveira AG, Reis AM, Kalapothakis E, Mahecha GA, Hess RA, Oliveira CA (2006) Effects of 3-beta-diol, an androgen metabolite with intrinsic estrogen-like effects, in modulating the aquaporin-9 expression in the rat efferent ductules. Reprod Biol Endocrinol 4(51):1–10. doi:10.1186/1477-7827-4-51

Pineau C, Dupaix A, Jegou B (1999) The co-culture of Sertoli cells and germ cells: applications in toxicology. Toxicol In Vitro 13(4):513–520

Pineau C, Sharpe R, Saunders P, Gerard N, Jegou B (1990) Regulation of Sertoli cell inhibin production and of inhibin α-subunit mRNA levels by specific germ cell types. Mol Cell Endocrinol 72(1):13–22

Piroli GG, Grillo CA, Hoskin EK, Znamensky V, Katz EB, Milner TA, McEwen BS, Charron MJ, Reagan LP (2002) Peripheral glucose administration stimulates the translocation of GLUT8 glucose transporter to the endoplasmic reticulum in the rat hippocampus. J Comp Neurol 452(2):103–114. doi:10.1002/cne.10368

Pollanen P, Soder O, Parvinen M (1989) Interleukin-1 alpha stimulation of spermatogonial proliferation in vivo. Reprod Fertil Dev 1(1):85–87

Rambourg A, Clermont Y, Hermo L (1979) Three-dimensional architecture of the golgi apparatus in Sertoli cells of the rat. Am J Anat 154(4):455–476. doi:10.1002/aja.1001540402

Rambourg A, Clermont Y, Marraud A (1974) Three-dimensional structure of the osmium-impregnated Golgi apparatus as seen in the high voltage electron microscope. Am J Anat 140(1):27–45. doi:10.1002/aja.1001400103

Rato L, Alves MG, Dias TR, Lopes G, Cavaco JE, Socorro S, Oliveira PF (2013) High-energy diets may induce a pre-diabetic state altering testicular glycolytic metabolic profile and male reproductive parameters. Andrology 1(3):495–504

Rato L, Alves MG, Socorro S, Carvalho RA, Cavaco JE, Oliveira PF (2012) Metabolic modulation induced by oestradiol and DHT in immature rat Sertoli cells cultured in vitro. Biosci Rep 32(1):61–69. doi:10.1042/bsr20110030

Rato L, Alves MG, Socorro S, Duarte AI, Cavaco JE, Oliveira PF (2012) Metabolic regulation is important for spermatogenesis. Nat Rev Urol 9(6):330–338

Rato L, Duarte AI, Tomás GD, Santos MS, Moreira PI, Socorro S, Cavaco JE, Alves MG, Oliveira PF (2014) Pre-diabetes alters testicular PGC1-α/SIRT3 axis modulating mitochondrial bioenergetics and oxidative stress. Biochim Biophys Acta 3:335–344

Rato L, Socorro S, Cavaco JE, Oliveira PF (2010) Tubular fluid secretion in the seminiferous epithelium: ion transporters and aquaporins in Sertoli cells. J Membr Biol 236(2):215–224. doi:10.1007/s00232-010-9294-x

Raychoudhury SS, Flowers AF, Millette CF, Finlay MF (2000) Toxic effects of polychlorinated biphenyls on cultured rat Sertoli cells. J Androl 21(6):964–973

Reagan LP, Gorovits N, Hoskin EK, Alves SE, Katz EB, Grillo CA, Piroli GG, McEwen BS, Charron MJ (2001) Localization and regulation of GLUTx1 glucose transporter in the hippocampus of streptozotocin diabetic rats. Proc Natl Acad Sci U S A 98(5):2820–2825. doi:10.1073/pnas.051629798

Regueira M, Riera MF, Galardo MN, Pellizzari EH, Cigorraga SB, Meroni SB (2014) Activation of PPAR alpha and PPAR beta/delta regulates Sertoli cell metabolism. Mol Cell Endocrinol 382(1):271–281. doi:10.1016/j.mce.2013.10.006

Reiter RJ, Rosales-Corral SA, Manchester LC, Tan DX (2013) Peripheral reproductive organ health and melatonin: ready for prime time. Int J Mol Sci 14(4):7231–7272. doi:10.3390/ijms14047231

Revelli A, Massobrio M, Tesarik J (1998) Nongenomic actions of steroid hormones in reproductive tissues. Endocr Rev 19(1):3–17

Ribbert H (1904) Die Abscheidung intravenos injizierten gelosten Karmins in den Geweben. Zeitschrift für Allgemeine Physiologie 4(15):201–214

Riera MF, Galardo MN, Pellizzari EH, Meroni SB, Cigorraga SB (2007) Participation of phosphatidyl inositol 3-kinase/protein kinase B and ERK1/2 pathways in interleukin-1beta stimulation of lactate production in Sertoli cells. Reproduction 133(4):763–773. doi:10.1530/rep.1.01091

Riera MF, Galardo MN, Pellizzari EH, Meroni SB, Cigorraga SB (2009) Molecular mechanisms involved in Sertoli cell adaptation to glucose deprivation. Am J Physiol 297(4):E907–E914. doi:10.1152/ajpendo.00235.2009

Riera MF, Meroni SB, Gomez GE, Schteingart HF, Pellizzari EH, Cigorraga SB (2001) Regulation of lactate production by FSH, iL1beta, and TNFalpha in rat Sertoli cells. Gen Comp Endocrinol 122(1):88–97. doi:10.1006/gcen.2001.7619

Riera MF, Meroni SB, Pellizzari EH, Cigorraga SB (2003) Assessment of the roles of mitogen-activated protein kinase and phosphatidyl inositol 3-kinase/protein kinase B pathways in the basic fibroblast growth factor regulation of Sertoli cell function. J Mol Endocrinol 31(2):279–289

Riera MF, Meroni SB, Schteingart HF, Pellizzari EH, Cigorraga SB (2002) Regulation of lactate production and glucose transport as well as of glucose transporter 1 and lactate dehydrogenase A mRNA levels by basic fibroblast growth factor in rat Sertoli cells. J Endocrinol 173(2):335–343

Rigau T, Rivera M, Palomo M, Fernandez-Novell J, Mogas T, Ballester J, Peña A, Otaegui P, Guinovart J, Rodriguez-Gil J (2002) Differential effects of glucose and fructose on hexose metabolism in dog spermatozoa. Reproduction 123(4):579–591

Robaire B, Viger RS (1995) Regulation of epididymal epithelial cell functions. Biol Reprod 52(2):226–236

Roberts KP, Zirkin BR (1991) Androgen regulation of spermatogenesis in the rat. Ann N Y Acad Sci 637:90–106

Robinson R, Fritz IB (1979) Myoinositol biosynthesis by Sertoli cells, and levels of myoinositol biosynthetic enzymes in testis and epididymis. Can J Biochem 57(6):962–967

Robinson R, Fritz IB (1981) Metabolism of glucose by Sertoli cells in culture. Biol Reprod 24(5):1032–1041

Rocha CS, Martins AD, Rato L, Silva BM, Oliveira PF, Alves MG (2014) Melatonin alters the glycolytic profile of Sertoli cells: implications for male fertility. Mol Human Reprod. doi:10.1093/molehr/gau1080. 10.1093/molehr/gau080

Rodriguez I, Ody C, Araki K, Garcia I, Vassalli P (1997) An early and massive wave of germinal cell apoptosis is required for the development of functional spermatogenesis. EMBO J 16(9):2262–2270

Rodriguez-Sosa JR, Dobrinski I (2009) Recent developments in testis tissue xenografting. Reproduction 138 (2):187-194. REP-09-0012. doi:10.1530/REP-09-0012

Roos A, Boron WF (1981) Intracellular pH. Physiol Rev 61(2):296–434

Roosen-Runge EC (1952) Kinetics of spermatogenesis in mammals. Ann N Y Acad Sci 55(4):574–584

Rosselli M, Skinner MK (1992) Developmental regulation of Sertoli cell aromatase activity and plasminogen activator production by hormones, retinoids and the testicular paracrine factor. PModS Biol Reprod 46(4):586–594

Ruiz-Pesini E, Díez-Sánchez C, López-Pérez MJ, Enriquez JA (2007) The role of the mitochondrion in sperm function: is there a place for oxidative phosphorylation or is this a purely glycolytic process? Curr Top Dev Biol 77:3–19

Russell L (1977) Desmosome-like junctions between Sertoli and germ cells in the rat testis. Am J Anat 148(3):301–312. doi:10.1002/aja.1001480302

Russell L (1977) Observations on rat Sertoli ectoplasmic ('junctional') specializations in their association with germ cells of the rat testis. Tissue Cell 9(3):475–498

Russell L, Clermont Y (1976) Anchoring device between Sertoli cells and late spermatids in rat seminiferous tubules. Anat Rec 185(3):259–278. doi:10.1002/ar.1091850302

Russell LD (1979) Observations on the inter-relationships of Sertoli cells at the level of the blood- testis barrier: evidence for formation and resorption of Sertoli-Sertoli tubulobulbar complexes during the spermatogenic cycle of the rat. Am J Anat 155(2):259–279. doi:10.1002/aja.1001550208

Russell LD (1979) Spermatid-sertoli tubulobulbar complexes as devices for elimination of cytoplasm from the head region of late spermatids of the rat. Anat Rec 194(2):233–246

Russell LD (1993) Form, dimensions, and cytology of mammalian Sertoli cells. In: Russell LD, Griswold MD (eds) The Sertoli Cell. Cache River Press, Clearwater, Florida, pp 1–37

Russell LD (1993) Morphological and functional evidence for Sertoli-Germ cell relationships. In: Russell LD, Griswold MD (eds) The Sertoli Cell. Cache River Press, Clearwater, pp 365–390

Russell LD, Brinster RL (1996) Ultrastructural observations of spermatogenesis following transplantation of rat testis cells into mouse seminiferous tubules. J Androl 17(6):615–627

Russell LD, Ettlin RA, Hikim APS, Clegg ED (1993) Histological and histopathological evaluation of the testis. Int J Androl 16(1):83

Russell LD, Franca LR, Brinster RL (1996) Ultrastructural observations of spermatogenesis in mice resulting from transplantation of mouse spermatogonia. J Androl 17(6):603–614

Russell LD, Peterson RN (1984) Determination of the elongate spermatid-Sertoli cell ratio in various mammals. J Reprod Fertil 70(2):635–641

Russell LD, Ren HP, Sinha Hikim I, Schulze W, Sinha Hikim AP (1990) A comparative study in twelve mammalian species of volume densities, volumes, and numerical densities of selected testis components, emphasizing those related to the Sertoli cell. Am J Anat 188(1):21–30. doi:10.1002/aja.1001880104

Russell LD, Steinberger A (1989) Sertoli cells in culture: views from the perspectives of an in vivoist and an in vitroist. Biol Reprod 41(4):571–577

Russell LD, Tallon-Doran M, Weber JE, Wong V, Peterson RN (1983) Three-dimensional reconstruction of a rat stage V Sertoli cell: III. A study of specific cellular relationships. Am J Anat 167(2):181–192. doi:10.1002/aja.1001670204

Ruwanpura SM, McLachlan RI, Stanton PG, Loveland KL, Meachem SJ (2008) Pathways involved in testicular germ cell apoptosis in immature rats after FSH suppression. J Endocrinol 197(1):35–43

Saether T, Tran TN, Rootwelt H, Christophersen BO, Haugen TB (2003) Expression and regulation of delta5-desaturase, delta6-desaturase, stearoyl-coenzyme A (CoA) desaturase 1, and stearoyl-CoA desaturase 2 in rat testis. Biol Reprod 69(1):117–124. doi:10.1095/biolreprod.102.014035

Sailer BL, Sarkar LJ, Bjordahl JA, Jost LK, Evenson DP (1997) Effects of heat stress on mouse testicular cells and sperm chromatin structure. J Androl 18(3):294–301

Sakai Y, Yamashina S, Furudate S (2004) Developmental delay and unstable state of the testes in the rdw rat with congenital hypothyroidism. Dev Growth Differ 46(4):327–334. doi:10.1111/j.1440-169x.2004.00748.x

Samih N, Hovsepian S, Aouani A, Lombardo D, Fayet G (2000) Glut-1 translocation in FRTL-5 thyroid cells: role of phosphatidylinositol 3-kinase and N-glycosylation. Endocrinology 141(11):4146–4155. doi:10.1210/endo.141.11.7793

Saunders P (2002) Germ cell-somatic cell interactions during spermatogenesis. Reproduction (Cambridge, England) Supplement 61:91-101

Saunders PT, Millar MR, Macpherson S, Irvine DS, Groome NP, Evans LR, Sharpe RM, Scobie GA (2002) ERbeta1 and the ERbeta2 splice variant (ERbetacx/beta2) are expressed in distinct cell populations in the adult human testis. J Clin Endocrinol Metab 87(6):2706–2715

Saunders PT, Sharpe RM, Williams K, Macpherson S, Urquart H, Irvine DS, Millar MR (2001) Differential expression of oestrogen receptor alpha and beta proteins in the testes and male reproductive system of human and non-human primates. Mol Human Reprod 7(3):227–236

Scarabelli L, Caviglia D, Bottazzi C, Palmero S (2003) Prolactin effect on pre-pubertal Sertoli cell proliferation and metabolism. J Endocrinol Invest 26(8):718–722

Scheepers A, Joost HG, Schurmann A (2004) The glucose transporter families SGLT and GLUT: molecular basis of normal and aberrant function. J Parent Ent Nutr 28(5):364–371

Schleicher G, Khan S, Nieschlag E (1989) Differentiation between androgen and estrogen receptor mediated effects of testosterone on FSH using androgen receptor deficient (Tfm) and normal mice. J Steroid Biochem 33(1):49–51

Schulz RW, Miura T (2002) Spermatogenesis and its endocrine regulation. Fish Physiol Biochem 26(1):43–56

Sertoli E (1865) Dell'esistenza di particolari cellule ramificate nei canalicoli seminiferi del testicolo unamo. Morgagni 7:31–40

Setchell B, Waites G (1975) The blood–testis barrier. In: Greep RO (ed) Handbook of physiology, vol 5, sect 7. American Physiology Society, Washington DC, pp 143–172

Setchell BP (1969) Do Sertoli cells secrete fluid into the seminiferous tubules? J Reprod Fertil 19(2):391–392

Setchell BP (1970) The secretion of fluid by the testes of rats, rams and goats with some observations on the effect of age, cryptorchidism and hypophysectomy. J Reprod Fertil 23(1):79–85

Setchell BP (1980) The functional significance of the bloob-testis barrier. Andrology 1:3–10

Setchell BP (2004) Hormones: what the testis really sees. Reprod Fertil Dev 16(5):535–545. doi:10.10371/rd03048

Setchell BP (2009) Blood-testis barrier, junctional and transport proteins and spermatogenesis. In: Molecular mechanisms in spermatogenesis. Springer, Berlin, pp 212–233

Setchell BP, Voglmayr JK, Waites GM (1969) A blood-testis barrier restricting passage from blood into rete testis fluid but not into lymph. J Physiol 200(1):73–85

Sharpe R (1994) Regulation of spermatogenesis. Physiol Reprod 1:1363–1434

Sharpe RM, McKinnell C, Kivlin C, Fisher JS (2003) Proliferation and functional maturation of Sertoli cells, and their relevance to disorders of testis function in adulthood. Reproduction 125(6):769–784

Shubhada S, Glinz M, Lamb DJ (1993) Sertoli cell secreted growth factor cellular origin, paracrine and endocrine regulation of secretion. J Androl 14(2):99–109

Sigillo F, Pernod G, Kolodie L, Benahmed M, Le Magueresse-Battistoni B (1998) Residual bodies stimulate rat Sertoli cell plasminogen activator activity. Biochem Biophys Res Commun 250(1):59–62

Sikka SC, Wang R (2008) Endocrine disruptors and estrogenic effects on male reproductive axis. Asian J Androl 10(1):134–145

Silber SJ (1978) Vasectomy and vasectomy reversal. Fertil Steril 29(2):125–140

Silva FR, Leite LD, Barreto KP, D'Agostini C, Zamoner A (2001) Effect of 3,5,3′-triiodo-L-thyronine on amino acid accumulation and membrane potential in Sertoli cells of the rat testis. Life Sci 69(8):977–986

Simões VL, Alves MG, Martins AD, Dias TR, Rato L, Socorro S, Oliveira PF (2013) Regulation of apoptotic signaling pathways by 5α-dihydrotestosterone and 17β-estradiol in immature rat sertoli cells. J Steroid Biochem Mol Biol 135:15–23

Simorangkir D, De Kretser DM, Wreford N (1995) Increased numbers of Sertoli and germ cells in adult rat testes induced by synergistic action of transient neonatal hypothyroidism and neonatal hemicastration. J Reprod Fertil 104(2):207–213

Singh J, Handelsman DJ (1996) The effects of recombinant FSH on testosterone-induced spermatogenesis in gonadotrophin-deficient (hpg) mice. J Androl 17(4):382–393

Siu ER, Mruk DD, Porto CS, Cheng CY (2009) Cadmium-induced testicular injury. Toxicol Appl Pharmacol 238(3):240–249. doi:10.1016/j.taap.2009.01.028

Siu MK, Lee WM, Cheng CY (2003) The interplay of collagen IV, tumor necrosis factor-α, gelatinase B (matrix metalloprotease-9), and tissue inhibitor of metalloproteases-1 in the basal lamina regulates Sertoli cell-tight junction dynamics in the rat testis. Endocrinology 144(1):371–387

Siu MK, C-h Wong, Lee WM, Cheng CY (2005) Sertoli-germ cell anchoring junction dynamics in the testis are regulated by an interplay of lipid and protein kinases. J Biol Chem 280(26):25029–25047

Skakkebaek NE, Rajpert-De Meyts E, Main KM (2001) Testicular dysgenesis syndrome: an increasingly common developmental disorder with environmental aspects. Hum Reprod 16(5):972–978

Skaper S (2012) The neurotrophin family of neurotrophic factors: an overview. In: Skaper SD (ed) Neurotrophic factors, vol 846. Methods in molecular biology. Humana Press, New York, pp 1–12. doi:10.1007/978-1-61779-536-7_1

Skinner M (1993) Secretion of growth factors and other regulatory factors. In: Russell LD, Griswold MD (eds) The Sertoli cell. Cache River Press, Clearwater, pp 237–248

Skinner M (1993) Sertoli cell-peritubular myoid cell interactions. In: Russell LD, Griswold MD (eds) The Sertoli cell. Cache River Press, Clearwater, pp 477–484

Skinner MK (2005) Sertoli Cell Secreted Regulatory Factors. In: Skinner MK, Griswold MD (eds) Sertoli cell biology. Elsevier Academic Press, New York, pp 107–120

Skinner MK, Fritz IB (1985) Testicular peritubular cells secrete a protein under androgen control that modulates Sertoli cell functions. Proc Natl Acad Sci U S A 82(1):114–118

Skinner MK, Fritz IB (1986) Identification of a non-mitogenic paracrine factor involved in mesenchymal-epithelial cell interactions between testicular peritubular cells and Sertoli cells. Mol Cell Endocrinol 44(1):85–97

Skinner MK, Griswold MD (1980) Sertoli cells synthesize and secrete transferrin-like protein. J Biol Chem 255(20):9523–9525

Skinner MK, McLachlan RI, Bremner WJ (1989) Stimulation of Sertoli cell inhibin secretion by the testicular paracrine factor PModS. Mol Cell Endocrinol 66(2):239–249

Slaughter GR, Means AR (1983) Follicle-stimulating hormone activation of glycogen phosphorylase in the Sertoli cell-enriched rat testis. Endocrinology 113(4):1476–1485. doi:10.1210/endo-113-4-1476

Smith LB, Walker WH (2014) The regulation of spermatogenesis by androgens. In: Seminars in cell and developmental biology. Elsevier, London

Sofikitis N, Giotitsas N, Tsounapi P, Baltogiannis D, Giannakis D, Pardalidis N (2008) Hormonal regulation of spermatogenesis and spermiogenesis. J Steroid Biochem Mol Biol 109(3):323–330

Sprando R, Russell L (1988) Spermiogenesis in the bluegill (*Lepomis macrochirus*): a study of cytoplasmic events including cell volume changes and cytoplasmic elimination. J Morphol 198(2):165–177

Stallard BJ, Griswold MD (1990) Germ cell regulation of Sertoli cell transferrin mRNA levels. Mol Endocrinol 4(3):393–401

Storey BT (2008) Mammalian sperm metabolism: oxygen and sugar, friend and foe. Int J Dev Biol 52(5):427–437

Su L, Mruk DD, Cheng CY (2011) Drug transporters, the blood-testis barrier, and spermatogenesis. J Endocrinol 208 (3):207-223. JOE-10-0363. doi:10.1677/JOE-10-0363

Sultan C, Paris F, Terouanne B, Balaguer P, Georget V, Poujol N, Jeandel C, Lumbroso S, Nicolas JC (2001) Disorders linked to insufficient androgen action in male children. Hum Reprod Update 7(3):314–322

Sun Y-T, Irby DC, Robertson DM, De Kretser DM (1989) The effects of exogenously administered testosterone on spermatogenesis in intact and hypophysectomized rats. Endocrinology 125(2):1000–1010

Syed V, Hecht NB (2001) Selective loss of Sertoli cell and germ cell function leads to a disruption in sertoli cell-germ cell communication during aging in the Brown Norway rat. Biol Reprod 64(1):107–112

Syed V, Soder O, Arver S, Lindh M, Khan S, Ritzen EM (1988) Ontogeny and cellular origin of an interleukin-1-like factor in the reproductive tract of the male rat. Int J Androl 11(5):437–447

Sylvester SR (1993) Secretion of transport and binding proteins. In: Russell LD, Griswold MD (eds) The Sertoli cell. Cache River Press, Clearwater, pp 201–216

Tabak AG, Herder C, Rathmann W, Brunner EJ, Kivimäki M (2012) Prediabetes: a high-risk state for diabetes development. Lancet 379(9833):2279–2290

Tapanainen JS, Aittomaki K, Min J, Vaskivuo T, Huhtaniemi IT (1997) Men homozygous for an inactivating mutation of the follicle-stimulating hormone (FSH) receptor gene present variable suppression of spermatogenesis and fertility. Nat Genet 15(2):205–206. doi:10.1038/ng0297-205

Taylor AH, Al-Azzawi F (2000) Immunolocalisation of oestrogen receptor beta in human tissues. J Mol Endocrinol 24(1):145–155

Thysen B, Morris PL, Gatz M, Bloch E (1990) The effect of mono(2-ethylhexyl) phthalate on Sertoli cell transferrin secretion in vitro. Toxicol Appl Pharmacol 106(1):154–157

Toyama Y, Maekawa M, Yuasa S (2003) Ectoplasmic specializations in the Sertoli cell: new vistas based on genetic defects and testicular toxicology. Anat Sci int 78(1):1–16

Toyama Y, Ohkawa M, Oku R, Maekawa M, Yuasa S (2001) Neonatally administered diethylstilbestrol retards the development of the blood-testis barrier in the rat. J Androl 22(3):413–423

Travis AJ, Jorgez CJ, Merdiushev T, Jones BH, Dess DM, Diaz-Cueto L, Storey BT, Kopf GS, Moss SB (2001) Functional relationships between capacitation-dependent cell signaling and compartmentalized metabolic pathways in murine spermatozoa. J Biol Chem 276(10):7630–7636. doi:10.1074/jbc.M006217200

Travis AJ, Tutuncu L, Jorgez CJ, Ord TS, Jones BH, Kopf GS, Williams CJ (2004) Requirements for glucose beyond sperm capacitation during in vitro fertilization in the mouse. Biol Reprod 71(1):139–145. doi:10.1095/biolreprod.103.025809

Trejo R, Valadez-Salazar A, Delhumeau G (1995) Effects of quercetin on rat testis aerobic glycolysis. Can J Physiol Pharmacol 73(11):1605–1615

Tsuruta JK, Eddy E, O'Brien DA (2000) Insulin-like growth factor-II/cation-independent mannose 6-phosphate receptor mediates paracrine interactions during spermatogonial development. Biol Reprod 63(4):1006–1013

Tsuruta JK, O'Brien DA (1995) Sertoli cell-spermatogenic cell interaction: the insulin-like growth factor-II/cation-independent mannose 6-phosphate receptor mediates changes in spermatogenic cell gene expression in mice. Biol Reprod 53(6):1454–1464

Tsuruta JK, O'Brien DA, Griswold MD (1993) Sertoli cell and germ cell cystatin C: stage-dependent expression of two distinct messenger ribonucleic acid transcripts in rat testes. Biol Reprod 49(5):1045–1054. doi:10.1095/biolreprod49.5.1045

Ulisse S, Jannini EA, Pepe M, De Matteis S, D'Armiento M (1992) Thyroid hormone stimulates glucose transport and GLUT1 mRNA in rat Sertoli cells. Mol Cell Endocrinol 87(1–3):131–137

Upadhyay R, Kumar A, Ganeshan M, Balasinor N (2012) Tubulobulbar complex: cytoskeletal remodeling to release spermatozoa. Reprod Biol Endocrinol 10(1):27

van Haaster LH, de Jong FH, Docter R, De Rooij DG (1992) The effect of hypothyroidism on Sertoli cell proliferation and differentiation and hormone levels during testicular development in the rat. Endocrinology 131(3):1574–1576

van Haaster LH, de Jong FH, Docter R, de Rooij DG (1993) High neonatal triiodothyronine levels reduce the period of Sertoli cell proliferation and accelerate tubular lumen formation in the rat testis, and increase serum inhibin levels. Endocrinology 133(2):755–760

Van Pelt A, de Rooij DG (1990) Synchronization of the seminiferous epithelium after vitamin A replacement in vitamin A-deficient mice. Biol Reprod 43(3):363–367

Verhoeven G, Willems A, Denolet E, Swinnen JV, De Gendt K (2010) Androgens and spermatogenesis: lessons from transgenic mouse models. Philos Trans R Soc B Biol Sci 365(1546):1537–1556

Villarroel-Espíndola F, Maldonado R, Mancilla H, vander Stelt K, Acuña AI, Covarrubias A, López C, Angulo C, Castro MA, Carlos Slebe J (2013) Muscle glycogen synthase isoform is responsible for testicular glycogen synthesis: Glycogen overproduction induces apoptosis in male germ cells. J Cell Biochem 114(7):1653–1664

Vogl A, Soucy L (1985) Arrangement and possible function of actin filament bundles in ectoplasmic specializations of ground squirrel Sertoli cells. J Cell Biol 100(3):814–825

Vogl AW, Pfeiffer DC, Mulholland D, Kimel G, Guttman J (2000) Unique and multifunctional adhesion junctions in the testis: ectoplasmic specializations. Arch Histol Cytol 63(1):1–15

Vogl AW, Pfeiffer DC, Redenbach DM, Grove BD (1993) Sertoli cell cytoskeleton. In: Russell LD, Griswold MD (eds) The Sertoli cell. Cache River Press, Clearwater, pp 39–86

Vogl AW, Young J, Du M (2012) New insights into roles of tubulobulbar complexes in sperm release and turnover of blood-testis barrier. Int Rev Cell Mol Biol 303:319–355

Voglmayr JK, Waites GM, Setchell BP (1966) Studies on spermatozoa and fluid collected directly from the testis of the conscious ram. Nature 210(5038):861–863

Wagner MS, Wajner SM, Maia AL (2008) The role of thyroid hormone in testicular development and function. J Endocrinol 199(3):351–365

Walker WH (2003) Molecular mechanisms controlling Sertoli cell proliferation and differentiation. Endocrinology 144(9):3719–3721

Walker WH, Cheng J (2005) FSH and testosterone signaling in Sertoli cells. Reproduction 130 (1):15-28. 130/1/15 [pii]. doi:10.1530/rep.1.00358

Wang RS, Yeh S, Chen LM, Lin HY, Zhang C, Ni J, Wu CC, di Sant'Agnese PA, deMesy-Bentley KL, Tzeng CR, Chang C (2006) Androgen receptor in sertoli cell is essential for germ cell nursery and junctional complex formation in mouse testes. Endocrinology 147(12):5624–5633. doi:10.1210/en.2006-0138

Wassermann GF, Bloch LM, Grillo ML, Silva FR, Loss ES, McConnell LL (1992) Biochemical factors involved in the FSH action on amino acid transport in immature rat testes. Horm Metab Res 24(6):276–279. doi:10.1055/s-2007-1003312

Watrin F, Scotto L, Assoian R, Wolgemuth D (1991) Cell lineage specificity of expression of the murine transforming growth factor beta 3 and transforming growth factor beta 1 genes. Cell Growth Differ 2(2):77–83

Weber JE, Russell LD, Wong V, Peterson RN (1983) Three-dimensional reconstruction of a rat stage V Sertoli cell: II. Morphometry of Sertoli-Sertoli and Sertoli–germ-cell relationships. Am J Anat 167(2):163–179. doi:10.1002/aja.1001670203

Welch JE, Brown PL, O'Brien DA, Magyar PL, Bunch DO, Mori C, Eddy EM (2000) Human glyceraldehyde 3-phosphate dehydrogenase-2 gene is expressed specifically in spermatogenic cells. J Androl 21(2):328–338

Welsh M, Saunders PT, Atanassova N, Sharpe RM, Smith LB (2009) Androgen action via testicular peritubular myoid cells is essential for male fertility. FASEB J 23(12):4218–4230

Wenger RH, Katschinski DM (2005) The hypoxic testis and post-meiotic expression of PAS domain proteins. Semin Cell Dev Biol 16(4–5):547–553. doi:10.1016/j.semcdb.2005.03.008

Williams AC, Ford WC (2001) The role of glucose in supporting motility and capacitation in human spermatozoa. J Androl 22(4):680–695

Williams J, Foster P (1988) The production of lactate and pyruvate as sensitive indices of altered rat Sertoli cell function in vitro following the addition of various testicular toxicants. Toxicol Appl Pharmacol 94(1):160–170

Wine RN, Chapin RE (1999) Adhesion and signaling proteins spatiotemporally associated with spermiation in the rat. J Androl 20(2):198–213

Wong C-H, Cheng CY (2005) The blood-testis barrier: its biology, regulation, and physiological role in spermatogenesis. Curr Top Dev Biol 71:263–296

Wong EW, Mruk DD, Cheng CY (2008) Biology and regulation of ectoplasmic specialization, an atypical adherens junction type, in the testis. Biochim Biophys Acta 1778(3):692–708

Wong RW, Kwan RW, Mak PH, Mak KK, Sham MH, Chan SY (2000) Overexpression of epidermal growth factor induced hypospermatogenesis in transgenic mice. J Biol Chem 275(24):18297–18301. doi:10.1074/jbc.M001965200

Wong V, Russell LD (1983) Three-dimensional reconstruction of a rat stage V Sertoli cell: I. Methods, basic configuration, and dimensions. Am J Anat 167(2):143–161. doi:10.1002/aja.1001670202

Wu X, Freeze HH (2002) GLUT14, a duplicon of GLUT3, is specifically expressed in testis as alternative splice forms. Genomics 80(6):553–557

Xiong W, Wang H, Wu H, Chen Y, Han D (2009) Apoptotic spermatogenic cells can be energy sources for Sertoli cells. Reproduction 137(3):469–479

Yamamoto H, Ochiya T, Tamamushi S, Toriyama-Baba H, Takahama Y, Hirai K, Sasaki H, Sakamoto H, Saito I, Iwamoto T (2002) HST-1/FGF-4 gene activation induces spermatogenesis and prevents adriamycin-induced testicular toxicity. Oncogene 21(6):899–908

Yan HH, Cheng CY (2006) Laminin α 3 forms a complex with β3 and γ3 chains that serves as the ligand for α 6β1-integrin at the apical ectoplasmic specialization in adult rat testes. J Biol Chem 281(25):17286–17303

Ye SJ, Ying L, Ghosh S, de Franca LR, Russell LD (1993) Sertoli cell cycle: a re-examination of the structural changes during the cycle of the seminiferous epithelium of the rat. Anat Rec 237(2):187–198. doi:10.1002/ar.1092370206

Yin Y, Hawkins KL, Dewolf WC, Morgentaler A (1997) Heat stress causes testicular germ cell apoptosis in adult mice. J Androl 18(2):159–165

Yoshida S, Yi N, Nakagawa T (2007) Stem cell heterogeneity. Ann N Y Acad Sci 1120(1):47–58

Yoshinaga K, Nishikawa S, Ogawa M, Hayashi S, Kunisada T, Fujimoto T (1991) Role of c-kit in mouse spermatogenesis: identification of spermatogonia as a specific site of c-kit expression and function. Development 113(2):689–699

Zhang H, Yin Y, Wang G, Liu Z, Liu L, Sun F (2014) Interleukin-6 disrupts blood-testis barrier through inhibiting protein degradation or activating phosphorylated ERK in Sertoli cells. Sci Rep 4:1–7

Zhang Y, Wang S, Wang X, Liao S, Wu Y, Han C (2012) Endogenously produced FGF2 is essential for the survival and proliferation of cultured mouse spermatogonial stem cells. Cell Res 22(4):773–776

Zhou Q, Nie R, Prins GS, Saunders PT, Katzenellenbogen BS, Hess RA (2002) Localization of androgen and estrogen receptors in adult male mouse reproductive tract. J Androl 23(6):870–881

Zhu LJ, Cheng C, Phillips DM, Bardin CW (1994) The immunohistochemical localization of α2-macroglobulin in rat testes is consistent with its role in germ cell movement and spermiation. J Androl 15(6):575–582

Zini A, Agarwal A (2011) Sperm chromatin. Biological and clinical applications in male infertility and assisted reproduction. Springer, New York

Zivkovic D, Hadziselimovic F (2009) Development of Sertoli cells during mini-puberty in normal and cryptorchid testes. Urol Int 82(1):89–91

Zysk JR, Bushway AA, Whistler RL, Carlton WW (1975) Temporary sterility produced in male mice by 5-thio-D-glucose. J Reprod Fertil 45(1):69–72